D1255642

1

Tickle Your Amygdala

Written and Illustrated by
Neil Slade

Brain Books and Music
Denver, CO

Tickle Your Amygdala

ISBN 978-0-9796363-5-6

First Edition, March 2012

Printed in the USA by
Qualimage Printing
Lakewood, Colorado
303-988-1560
qualimage@msn.com

Many thanks to all of my friends, old and new, who gave of their generous time, energy, and wisdom towards the making of this book.

Contents

QUICK INTRODUCTION

PART ONE: *The Illustrated Tickle Your Amygdala* is a fun, simple, and quick explanation to help you get the general idea of what amygdala tickling is all about. Now you have something to share with that cute guy or girl sitting next to you on the bus or in the coffee shop when you can't think of anything else to say.

PART TWO: *The Amygdala Tickling Three Course Gourmet Presentation* includes 1) Detailed Explanations of Amygdala Tickling, 2) Amygdala Interviews, and 3) Ways to Do It. This includes the science behind amygdala tickling, stories, and 52 Ways to Tickle- That's a new way to turn on the best part of your brain for every week of the year.

The Amygdala Interviews are excerpts from fifty-six conversations I had with people from all over the world, many of them top experts in their chosen field. It is a diverse selection of people in many occupations and of many interests from the age of twenty to eight-five years old. The original conversations far exceeded one-hundred hours total, all spent in glial-ful conversation, with the most relevant portions presented here for your own frontal lobes pleasure.

Some of the people I spoke with are quite well known, but others you probably have not heard of before. Importantly, however, they all have the same basic mind motor that you have, and it fundamentally works in the very same way yours does.

The thread that ties all of these unique tales together is the story of how the human brain produces amazing results when the amygdala is tickled forward.

Having the ability to instantly tickle your own brain is having a lovely cake that magically reappears as soon as you think you've finished it off. What could be better?

The amygdala and what it does has been a non-secret since amygdala started appearing in the brains of mammals over 65 million years ago. All you have to do is pay attention to it. This book is just a post-it note reminder about something your parents and your first grade teacher probably never told you about way back in grade school when you were busy gawking at pictures of Tyrannosaurus Rex.

So then- onward to some 22nd Century brain magic...

-Neil Slade
March, 2012

PART ONE:

THE ILLUSTRATED

TICKLE
YOUR
AMYGDALA

Chapter 1
AMYGDALA TICKLING FUN-DA-MENTALS

Robert Schneider
Writer

RS: "...But anyway, one day I was driving along, thoroughly depressed, and I did a little amygdala click and became completely blissed out.

NS: (laughs) "You weren't taking any drugs, correct?"

RS: "No. No drugs whatsoever, but it was as if I had taken a very strong one- it was that big a change. That feeling persisted for a good six or eight months I guess. A permanent high. Every time I clicked forward I'd get on a big high. It was simply that flip, that simple little flip of the amygdala. This is what is so extraordinary to me, that it happened, and that there wasn't anything subtle about it. It was just a complete change of outlook."

Imagine you have a feather inside your brain.
Use it to directly tickle your brain's Pleasure Spot.

That's one way to Tickle Your Amygdala.

Do you sometimes feel like a dog waiting for scraps, looking up at the table while the rest of the family is scarfing down a big Thanksgiving dinner?

So, how are you going to get a fat piece of that pumpkin pie, eh?
Luck?
Fangs?
A giant super computer?

Well, it's your lucky day! You already have the exact tool to get what you really want and need-

You were born owning the most POWERFUL tool on Earth. It is the most complex and remarkable device yet that anyone has ever discovered. This is your very own-

HUMAN BRAIN

Save Money! Wipe your chin and forget about that vastly overpriced and environmentally unsound $5000 Super-Duper Liquid Cooled 100-Core Deluxe portable combo computer-cell phone-washing machine-can opener that you've been drooling over.

You ALREADY have the most amazing and powerful doo-dad ever created, and it is sitting right between your very own waxy two ears.

Fact: Your brain is an Infinity Mind Motor and Calculator with literally more possible connections than there are grains of sand on all of the beaches on Earth, more connections than there are stars in the sky on a clear dark night, or for that matter, you have more connections in your brain than the total number of elementary particles in the universe- as calculated by Dr. Carl Sagan, Dr. Richard Restak, and others.*

By comparison, any other machine is about as impressive as a lonely squirrel chewing on a stale wet watermelon seed.

*Dr. Carl Sagan, cosmologist, *Cosmos* (1980, 2002), *The Dragons of Eden* (Pulitzer Prize, 1977)
*Dr. Richard Restak, Neurologist, *The Brain* (1984), *Mysteries of The Mind* (2000)

BASIC TERMS OF USE: TICKLING YOUR AMYGDALA

So, what is the key to using this incredible brain machine that you have on your shoulders to figure out which way to the jackpot?

It is this:
Inside your personal super mind motor is a Master Compass that works like pure magic:

The Amygdala.

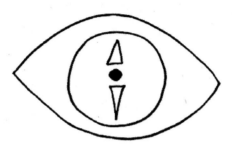

The amygdala is part of a brain circuit that quickly tells you which way to go- when you need to know it.

It tells you via emotional feedback how to know exactly what is bad for you and also what is good for you. This brain circuit computes:

Pleasurable Emotions as Reward
and
Unpleasant Emotions as Deterrent

Here is a photograph of a real human brain outside its container along with a couple of real amygdalae. I know this looks kind of gross, but I want to make sure readers of this book know that I am not a lunatic and just making all this stuff up.

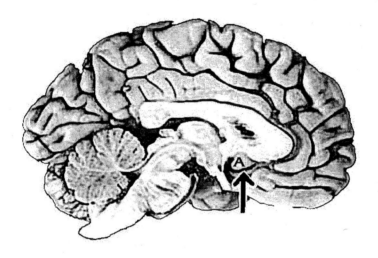

(Above: Cross section)

You have two amygdala in your brain, but they both pretty much do the same thing. So we just say the singular "amygdala" to refer to them both.

(Below: View from underneath)

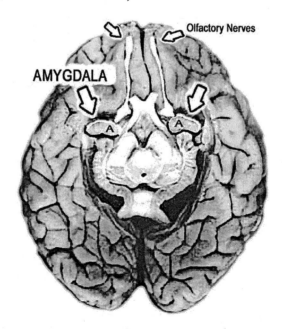

Dr. Robert Neumann
Head Neurosurgeon, University of Colorado Medical Center Hospital
"The concept that we each have far-reaching untapped potential is a very tempting concept, because the next question it then leads one to is then, 'How do I get to it?'"

Although the amygdala responds to external cues and external things like a nice kiss or the promise of a new toy unwrapped, you can also self-stimulate your own amygdala directly- and powerfully- by using your own internal thoughts and behaviors. This is called

TICKLING YOUR AMYGDALA

Tickling your amygdala means to observe and directly stimulate your brain's master Reward-Pleasure circuit.
You do it by using your own brain and thought processes.
Imagine that.

What could possibly be better than being able to directly tickle your own brain's pleasure circuits and your Feel Good spot?!

Tickling the amygdala is easier than scratching an itch, licking your favorite flavor ice cream cone, or eating chocolate cream pie.

It is like a dessert that never runs out- even when you don't have any money in your pocket.

You can tickle your amygdala any time and at any place.

Unlike scratching your butt, you don't have to worry about being embarrassed that you are doing it in front of somebody you are trying to impress.

As it turns out, tickling your amygdala also turns on your brain's built-in genius problem solving, creativity, and intelligence circuits. That's right- in the middle of your own cranium you have the very same internal circuitry that little Albert Einstein was born with, and which he later tickled himself to write his world famous equations.

Tickling your amygdala works better than anything else for *you* and for *everyone around you.*

Tickling your amygdala is as easy as flipping a light switch on, it costs nothing, and you can't be put in jail for doing it.

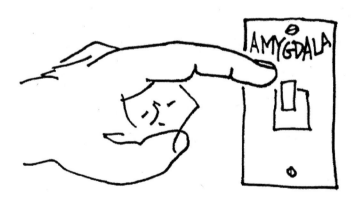

This all sounds great, but, it's just pretend, right?

Nope.

Between 1995 and 2009, Dr. Sarah Lazar, Dr. Herbert Benson and other researchers at the Harvard Medical School demonstrated conclusively in a number of research projects that one could instantly and positively stimulate one's amygdala with sheer thought. The results were demonstrated and recorded using functional MRI brain scanning machines and have been repeated many times by others in similar laboratory experiments.

People all over the world are now reporting their ability to tickle their amygdala. Any kind of people; Smart people, silly people, professionals, unemployed workers, skateboarders, scientists, gardeners, you name it.

People just like you and your mother.

Marie-Louise Oosthuysen
Brain Education Researcher
You have to stimulate that neuro-pathway between the amygdala and the pre-frontal cortex, so you can think things through very quickly. And the best way to do that is to tickle the amygdala forward.

Kyle Ridgeway
Physical Therapist, (Neurophysiology)
"Just the thought of that feather really engages your imagination, it's instant. It is so simple, but it's instantly effective."

But there is not just one way to tickle your amygdala.

You will discover how to tickle your amygdala in your very own way.

When you do- you'll know it:

You will smile- and even at times, fall over laughing, because it's so simple to do.

BRAIN RADAR

When you tickle your amygdala you automatically turn on Brain Radar.

Brain Radar is the application of "Whole Brain Power", a powerful combination of logic and reason combined with extraordinary intuitive perception.

Brain Radar provides you with seemingly "magical"- but completely *real* means for arriving at

The Right Place at The Right Time with The Right Solution

Elizabeth Slowley
Massage Therapist
"What's interesting is how fast it happened, and how powerful our thoughts are. Activating our higher brain power activates our whole body vibration."

Your built-in Brain Radar guidance system delivers you right on target.

Brain Radar is like always having a magic wand in your pocket- except that it is *not* make-believe. It comes from a real understanding of how your brain works and how to access the infinite potential in yourself.

Paul Epstein
Independent Record Store Owner
"I'm the master of my own destiny- as long as I'm not afraid to do things differently."

POPPING YOUR FRONTAL LOBES

Regular amygdala tickling will eventually allow you to: "Pop Your Frontal Lobes".

We are all familiar with popping chewing gum or popping your eardrums on a plane. Believe it or not- the brain is actually capable of more than that.

Popping Your Frontal Lobes is like hitting the *Brain* Jackpot.

It is the astonishing sudden peak "Eureka!" moment of great significance, discovery, solution, understanding, and overwhelmingly positive emotion that far eclipses normal experience. You are at last licking that big ice cream cone in the sky.

If you think about it for just a few seconds, you'll remember something like this *has* happened to you, or something quite close to it. Maybe it was just a little thing, or maybe it was a big thing.

Just think for a second and remember a time when suddenly all of the pieces fit together, when you finally got it, what you had been looking for, for all that time. You can Pop! your frontal lobes and make it happen again. And again, and again.

Shari Harter
S.F., CA
"I was in the bathtub one night, and I was in a relaxed self-contained place and I was doing my visualization there. That's how I would tickle my amygdala. I could stay focused and do that for quite a while, and then I had an inner Pop! and I recognized it."

You no longer have to tolerate every moment as if you are bored out of your skull, as if you are doomed to clean out your cat box for eternity.

Instead, you can choose to tickle your amygdala, turn on Brain Radar, and Pop Your Frontal Lobes *in a big way* and get that big piece of pie that you've been drooling over and that has been out of reach for so long.

Science has now demonstrated that the biggest and smartest part of your brain is also the most fun and pleasurable to use. It's right there waiting to be tickled by you, twenty-five hours every day.

It's as easy as... well... pie.

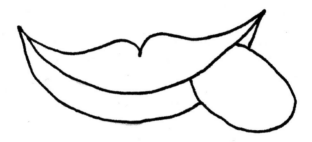

Anything else is just an accident.

Jim Casart
Certified Public Accountant
"Right there in my mirror, the answer to my problems is staring right back at me. This is a very empowering thought."

So, the first step is deciding a few things:
WHAT DO YOU WANT?

1..

2..

3..

4..

5..

You can solve your problems and get what you really want, as a result of your own easy and predictable method of tickling your amygdala instead of stumbling around in the dark with your fingers crossed.

Or if you prefer, you can put a sack over your head and continue to enjoy banging it against the wall until the cows come home.

Which do you prefer?

Important Security Notice: Tickling your amygdala can be done in complete and utter privacy: The CIA and FBI cannot listen to the phone calls that you make to yourself between your ears.

So, if your brain is so great, WHAT'S THE PROBLEM??

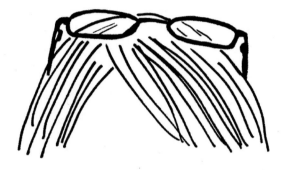

"Now where did I put my glasses?

Haven't you ever looked all over the house for something, only to finally find that it was under your nose the whole time?
Humans brains are funny that way.

Don't worry.
You can now tickle your amygdala and find what you've been missing all along.

Some people say, "You don't need your brain! You don't need to know anything about it!"

No need to think.

This is true.

In 1792 the French discovered that one could solve all possible problems by removing the brain and then throwing it away. This worked for many, especially those who found their brain to be a bothersome inconvenience and an inferior organ of the body.

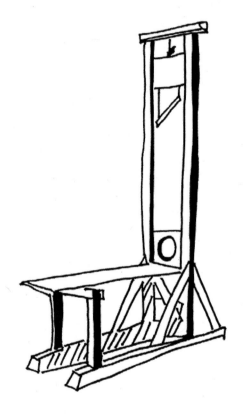

If you are one of those people, please give this book away to somebody else or use it in your compost pile.

However, if you are open to an alternative possibility, or the idea that your brain is somewhat useful, please keep reading.

Survival and Happiness means getting to Chocolate Cream Pie*.

*(Or whatever it is that you need to stay alive and find meaning. If you don't like chocolate, then banana or lemon custard is okay.)

L.I.F.E. is a
Line
Indicating
Forward
Energy

of YOU moving towards the pie or the pie moving towards you.

When L.I.F.E. works right and you are going in the right direction
It FEELS GOOD.

When things go wrong and you are going in the wrong direction
It FEELS BAD.

It feels good to go forward towards something good.

It feels bad to go hungry, towards zilch.

You can KNOW exactly
which DIRECTION you are going

Before you get there

by how you
FEEL INSIDE YOUR BRAIN

Inside YOUR BRAIN there Is A
MASTER COMPASS

This is Your
AMYGDALA

The amygdala is an emotional compass that tells you if you are going in the right direction.

The amygdala does this with EMOTIONS.

The amygdala is a little button inside the middle of your brain here:

    You actually have two of them, one for each side (hemisphere) of your brain, behind each eye and about one inch inside from the top of your ear.
    We call it "amygdala" (singular) because they both do basically the same thing, one for each hemisphere of your brain.

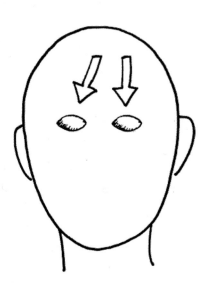

The amygdala is at the emotional HEART of your brain.

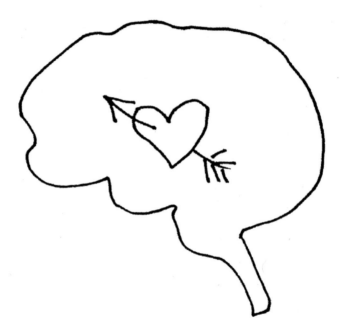

Your amygdala has memorized from experience and knowledge what Feels Good and what Is Good For You.

It has also learned what Feels Bad and what Is Bad For You.

It provides a lightning fast brain shortcut that works faster than you can say

"Gimme a piece of dat pie!"

Have you ever used a compass? If you have one around the house, go get it right now and have another look.

The amygdala uses EMOTIONAL MAGNETICS to indicate L.I.F.E. Direction towards or away from Chocolate Pie and other good things.

Did you ever wonder why some things make you feel good and why other things make you feel bad?

Here's the reason:
Positive-Pleasure emotions ATTRACT us to those things that keep us alive.
Negative-Displeasure emotions REPEL us away from things that threaten our survival.

This is just like how a compass needle is attracted to cute playful polar bear cubs in the North Pole, and points away from squawky penguins in the South Pole.

Your amygdala's FEELING BAD EMOTIONS show L.I.F.E. Direction away from PIE-

Amygdala gives you "YUCK!"

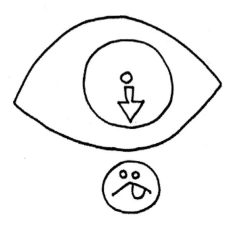

Negative Emotions REPEL us away from those things that harm us. Here is one example of how it works:

1) When you were little, you burned your finger on the stove and it FELT BAD.

2) Your amygdala learned that a lit stove burner + finger is not a healthy thing.

3) So, from that point on, your amygdala gives you FEAR when you sense your finger is too near a hot stove flame.

4) You are REPELLED by things that look too hot for your finger.

This means that you can avoid hot things without getting burned every time. Thus, your amygdala's Emotional Magnetics REPEL you from harmful looking things *before* you stick your finger in 'em. That's quite a handy thing.

This advance negative brain signal is called an Amygdala Bite.

Amygdala Bite is when the amygdala flashes a RED FEEL BAD SIGNAL.

STOP! DANGER A-HEAD!

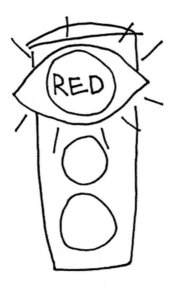

Forward L.I.F.E. Direction Automatically Puts On The Brakes with
An Amygdala Bite

When your direction is AWAY from Chocolate Cream Pie, L.I.F.E. flow is reduced. You are going in the wrong direction, your amygdala gets bitten, and you feel bad.

This is like being sent to your bedroom without dessert.

When your amygdala gets bitten you feel like you are going down the drain.

When something BITES an Amygdala, people have been known to:

EXPLODE

DRIVE A CAR OFF A CLIFF

HIDE UNDER THE BED

CRY A POOL OF TEARS

When your amygdala points down and backwards, L.I.F.E. sucks. This is like having no pie and you are still hungry.

Do you ever feel like your are going the wrong direction?

yes ☐   no ☐

Do you ever feel like your amygdala is getting bitten?

yes ☐   no ☐

Do you wish you had a dependable COMPASS inside your brain that you could follow that would help you go in the right direction, and keep from getting bitten?

yes ☐   no ☐

SURPRISE! SURPRISE!

You *do* have a compass inside your brain.

The only problem is that you may not always correctly be reading your compass. Maybe you don't even know you have a compass or that you have an amygdala.

That's all.

FEELING GOOD EMOTIONS show Energy Direction Towards PIE-
Amygdala gives you "YUM!"

Positive Emotions attract us towards those things that help us. Here is one example of how it works:

1) Mom smiles and then hugs and feeds you, and you FEEL GOOD.

2) Your amygdala learns Mom Smile + Hugs and Yummy Food is healthy.

3) So, from that point on, your amygdala gives you HAPPY when you see Mom smile.

4) You are then ATTRACTED by things that look like Mom smiling.

Thus, your amygdala's Emotional Magnetics automatically attract you towards helpful looking things. That's also very handy. And it feels good. Wowie Zowie!

This advance positive brain signal is an Amygdala Tickle.

Amygdala Tickle is when the amygdala shows a
GREEN FEEL GOOD SIGNAL

GO FORWARD!

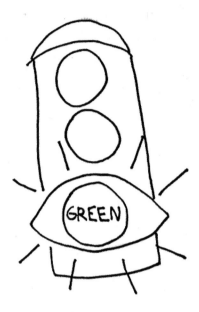

L.I.F.E. Direction Flows Forward-
Amygdala Tickle.

When your direction is towards Chocolate Cream Pie, L.I.F.E. is going in the right direction, you feel good, *and your amygdala is being tickled.*

Your compass is indicating you are going the right way.

In summary:

The overall Quality of Your Life depends upon what your Amygdala Compass says about your relative position to Chocolate Cream Pie.

To review quickly here:

If you are getting an **Amygdala Bite**, you are going in the wrong direction, and L.I.F.E. sucks.

If you are getting an **Amygdala Tickle**, you are going in the right direction, and L.I.F.E. is delicious.

NOTE: Other historical observations-

In 1687 Sir Isaac Newton wrote about this subject which he first expressed as his "4th Law of Motion". His original notes were only recently discovered on ketchup stained paper napkin stuck to the wall behind an old wooden bench in an old Cambridge pub newly under demolition, formerly known as "The Pig and The Poke". These misplaced notes were likely drawn up and then forgotten after Sir Newton had a few too many pints of ale under his belt. This is probably the reason for their absence in the original manuscript of "Mathematical Principles of Natural Philosophy".

In any case, Newton's long lost 4th Law of Motion states:

"For every good snack action there is an opposite and equal glee action."

Not to be outdone by his predecessor, few people today know that Albert Einstein addressed this same topic shortly after the publication of his 1905 paper "On the Electrodynamics of Moving Bodies". Einstein wrote his little known *third* theory after a particularly unpleasant spat with his first wife, Melva, who was constantly criticizing Albert for "…spending more time with that damn violin than you do with me."

On one particular night after a marital wrestling match, Einstein stormed out of the house. Instead of moping about the dark neighborhood streets he wandered into an all night Zurich café to think about something more appealing. On this, yet another night of miraculous inspiration during his "miracle year", he stared into the remaining crust pieces from a delicious apple turnover he had just finished and instantly drew up his observations in his follow up paper, "The *Really* Special Advanced Theory of Relativity" in which he states in simple relativistic elegance:

$$S = \pi r \bigcirc$$

(Smile equals
Pie Are Round)

BUT- Ah ha! You say-

"Not EVERYTHING that feels good is good for me!! How about eating lots and lots of candy?! That gives me a stomach ache!"

Of course it gives you a stomach ache! If it doesn't feel good, it's not good for you.

You have to consider *EVERYTHING*, the big picture. Not just what is immediately right in front of you.

When you learn how to *really* read your Amygdala Compass and then tickle your amygdala, you learn how to see more than one inch in front of your nose-

And then you don't get any more tummy aches.

Chapter 3

LITTLE BRAINS and BIG BRAINS

The amygdala is connected to many areas of your brain, like a central hub. This allows it engage with all kinds of physical and mental activities, both advanced and primitive, intellectual and rudimentary.

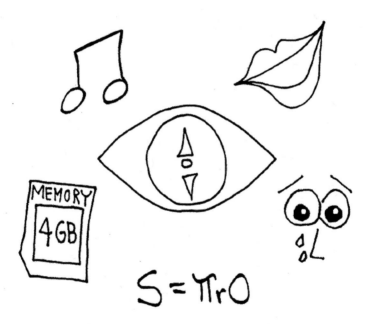

To understand how the amygdala fits into the overall picture of brain function we shall first examine three simple models of human brain physiology...

THE TRIUNE BRAIN

Dr. Paul MacLean, Director of the Laboratory of Brain Evolution and Behavior, National Institute of Mental Health (1971-1985) developed the triune (3 in 1) model of the human brain.

The triune model provides an overall quick impression of how the main parts of the brain are laid out. Like any model it is not perfect, but it is still useful to anyone wanting some general insight to how the human brain is constructed and how it works.

MacLean tells us that human brain is "three brains in one", built something like an apple: A reptile brain in the center like the inner-most seeds, surrounded by a more advanced mammal brain core, and further surrounded by a big delicious and juicy advanced primate brain and frontal lobes.

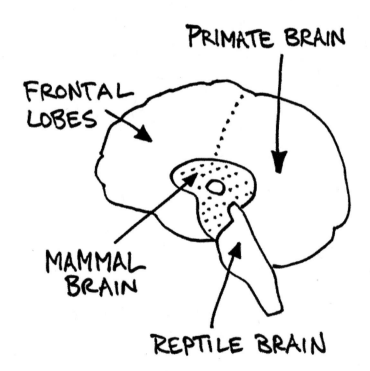

1) The inner **reptile brain** computes **basic body functions, non-thinking reactions, fight or flight, attack counter-attack response.**

When you stub your toe, grab it and yell "ARRRGGGGGG!" this is your reptile brain doing its thing. It's automatic. You don't think about it. You just do it.

When your twin sister Gladys grabs your hair at the dinner table and you automatically react by emptying your pudding on top of her head, this is both of your reptile brains at work.

Amygdala Bite. Ouch.

2) The surrounding **mammal brain** adds on **emotions and more complex social behaviors.**

When Gladys cries because she now has pudding flavored hair and yells "Mommy! Irving put pudding on my head! Do something!", it is because her mammal brain recognizes that this was not a very nice thing to do and it feels and expresses this as an emotion.

When Mommy sends both you and Gladys to your bedrooms without any more dessert, it is also her own motherly mammal brain at work. "OOHHH those kids are driving me Kkkeeeerrrraaaaaaaaazzzzzzyyyy!"

Amygdala Bite. Waaahhh!

3) The **primate brain and frontal lobes** adds to this **advanced thinking and perception, i.e. CICIL cooperation, imagination, creativity, intuition, logic.**

The glorious Frontal Lobes are the largest and most advanced part of the human brain. By itself, it makes up roughly a full 1/3 of the human brain, a far larger area than any other single part of the human brain.

When Gladys and you have forgotten all about pudding the next day and you work together gleefully and enthusiastically to design a two-foot tall explosive vomiting volcano for your school science fair project that later wins the blue ribbon first prize, it is your advanced primate brain and frontal lobes that did the trick.

Cooperating and creating your volcano together tickles both of your amygdalae! You win-win the race.
Big Amygdala Tickles.
Hurray! Yum.

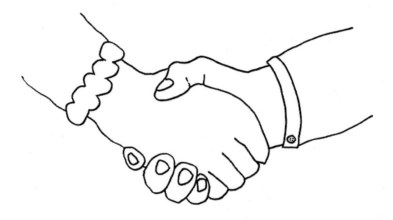

Each triune brain element can be looked at separately, but also as how they are interconnected.
All can be mixed and matched in an endless variety of ways, like the contents of your wardrobe.
Any day's outfit is made up of: Itchy or comfy underwear (reptile brain); bland or colorful shirts and pants (mammal brain); and fancy or plain overcoats (primate brain and frontal lobes).

When your outfit looks good to you- Your amygdala gets tickled.

The REACTIVE BRAIN and The REFLECTIVE BRAIN

Neurologist Richard Restak, M.D. has provided us with an even simpler way to look at the human brain: One part is "*automatic*" and reactive, and the other part is "*controlling*" and reflective.

Reactive Brain

The parts that are automatic-reactive are beyond our conscious control. The Reactive Brain would also include all of those structures lying underneath the cortex as MacLean described as the Mammal and reptile brain.

"Those regions that are involved with automatic activity are concentrated toward the back (occipital), top (parietal), and side (temporal) lobes."

Reflective Brain

The Controlling-Reflective Brain calculates by applying linear and verbal logic, non-verbal perception and intelligence, and higher intuition. We can also control and modify what The Reflective Brain does.

"Controlled processes, in contrast, occur mainly in the front (orbital and prefrontal) areas, with the prefrontal cortex especially important since it integrates information from all other parts of the brain, fashions long- and short-term goals, and directs our overall behaviors. Think of the frontal lobes as the CEO of the most complex organization in the world, the human brain." (R. Restak, *The Naked Brain*, 2006)

When your Reactive Brain and Reflective Brain work together in perfect balance and things go right, your amygdala gets tickled.

The TRY-AGAIN BRAIN MODEL

By combining MacLean's model with Restak's view, this book proposes a fabulous shiny new 22nd Century model of the human brain to supplement or stand as an alternative to the older Triune and Reactive/Reflective Brain models. This is The *Try-Again* model of the human brain.

The "Try-Again" model of brain consists of:

1) The Little Yucky Brain that computes Too Little or Too Much.

2) Goldilocks smack in the middle of things.

3) The Big Yummy Brain that computes Just Right.

Little Yucky Brain = Reactive Reptile Brain
Big Yummy Brain = Reflective Frontal Lobes

Goldilocks is smack in the middle of things, not always sure which way to go. (That's "you" and how your AMYGDALA feels).

When Goldilocks talks only to the Little Yucky Brain, all she gets is a chair that's either too big or too small, porridge that's either too hot or too cold, or a bed that's either too hard or too soft.

Amygdala Bite.

You see, Goldilocks is easily distracted by any old thing that comes her way. She has a poor memory, she is easily influenced by commercial advertising that she hears on the little radio inside her thatched cottage, and she further comes under constant peer pressure from her cousin Snow White and other bad influences such as The Big Bad Wolf and The Wicked Witch of the West. Using her Little Yucky Brain alone, she just doesn't know who her real friends are.

On the other hand, one day Goldilocks discovers she has a Big Smart Yummy Brain. She realizes that when she talks to her Big Yummy Brain she can then Try-Again and look for and find a chair, porridge, and a bed- and all of them are *Just Right.*

When Goldilocks finds that things are Just Right, her amygdala gets tickled.

LITTLE REPTILE BRAIN LIMITATIONS

Whatever you call it, The Reptile Brain, the Automatic Brain, the Reactive Brain, or the Little Yucky Brain- this is the part of the human brain that just reacts to what comes your way without thinking about it much. It is limited in what it can do with what it gets. When it sees lemons, The Little Brain is nothing but a sour puss.

Your reptile brain only reacts with what chance or bad luck delivers- good fortune or rotten luck. Using *nothing but* your reptile brain is like getting stuck with dry crackers for your birthday.

That Bites.

When you use just your reptile brain it's like only having one cylinder of your big a six cylinder car firing. You can't fix anything, you're just stuck reacting- "Arg%$#*@^#* a!!!!!"

You've seen this happen when your husband has a two-year old kiddie temper tantrum. He's only using crackers for brains.

Or it's as if you're trying to drive across town after your neighbor's kid had pulled off all but one of the spark plug wires on your car's engine.

Clunky slow going.

Your Little Reptile Brain cannot see "Outside the Box". It can only see and smell what is one inch in front of its nose only.

This is great if you have a piece of cheese in front of you. This is not so great if what you need is not in front of you.

Your Yucky Little Reptile Brain has no imagination, it can't plan anything, or understand time. Stuck in your Little Brain is like being stuck in the middle of the desert, without a locomotive or any way to get anywhere.

Your reptile brain cannot fathom cause and effect, nor understand, nor cooperate. That's good if you are about to get run over by a truck or you're being chased by a lion.

But otherwise, when your amygdala flashes RED and gets bitten, all your Yucky Brain can do is explode, drive off a cliff, cry, run, or bite back. Fight or Flight. No other options.

The Little Brain can only respond to whatever is at hand at that moment- Good or Bad, no more, no less. Giggly wiggly when it gets a cupcake- but Cry Baby Cry when it doesn't.

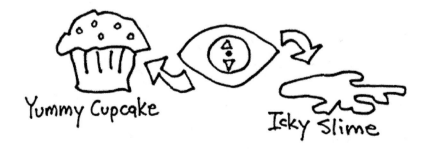

THE AMYGDALA CONNECTED

The amygdala sits right smack in the middle between your reptile brain and your Big Yummy Frontal Lobes.

Both parts of your brain are connected to your Amygdala Compass. And your reptile brain can read your amygdala without any problem...

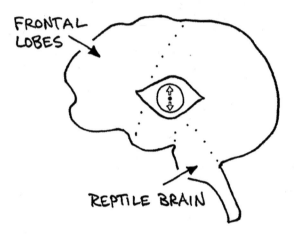

But, your little reptile brain can be a LITTLE DEVIL when it's reading your Amygdala Compass!

The Little Yucky Brain can temp you and say, "Hey Goldilocks! That sweet treat looks mighty good right NOW!"

But that treat can turn out to be a trick and a bag of worms tomorrow!

Is L.I.F.E. just a toss of the dice to which you can only react?

Half the time or more your Little Reptile Brain will have you driving down a one way Dead End Street with no reverse gear. And if The Little Brain is the only brain you're using, you never know it's the end of the road until it's too late.

HOWEVER FORTUNATELY-

Your amygdala is also connected to the MOST ADVANCED part of your brain- The BIGGEST single part of your brain,

THE FRONTAL LOBES

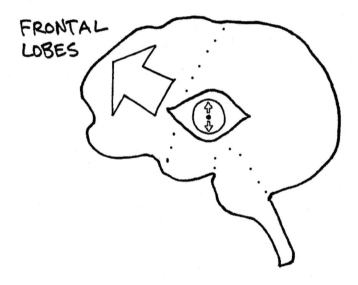

Your frontal lobes are flexible and nimble, yet powerful. It can make a quart of lemonade out of any bunch of lemons. It can solve any puzzle that comes your way.

You can tickle your amygdala at any time using your frontal lobes-
No matter what is happening outside of your brain, instead of being controlled by Stuff that Happens to Happen-

You can TICKLE YOUR AMYGDALA to feel GOOD.

Here's how it works...

Your frontal lobes can instantly change and rotate the direction your Amygdala Compass is pointing- From Bad Direction to Good Direction.

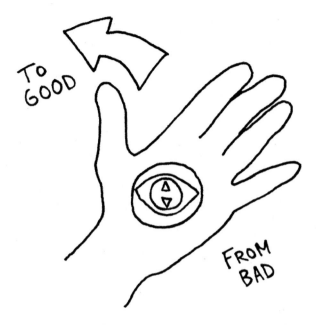

1) The amygdala is hardwired in the brain to BITE when you are going the wrong way, and to TICKLE when you are going the right way.

2) Your frontal lobes can see which direction your Amygdala Compass is pointing by evaluating your Emotional Magnetic Direction- positive or negative emotion.

3) If you are pointing the wrong way, your frontal lobes allows you to intelligently adjust your interior and exterior direction – no matter what else going on around you.

4) You re-align with the L.I.F.E. direction and face in the direction of survival and happiness and go that way... by changing YOUR direction.

You make lemonade out of lemons.
You see the glass half full.
You whistle while you work.
You follow your bliss.
You powder your armpits.
Your amygdala gets tickled.

Instead of just watching your amygdala twirl and spin at random like a tornado on a crummy one channel Reptile Brain TV set...

Your advanced frontal lobes allow you to tune into any station on the INFINITY DIAL. You can pick out a much better program any time you wish.

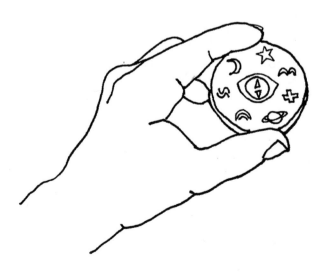

How can you steer your L.I.F.E. mobile and follow the direction of your Amygdala Master Compass so it points towards survival and happiness??

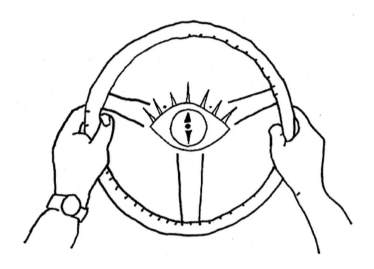

It's Easy:

C.I.C.I.L.

C.I.C.I.L.

Cooperation
Imagination
Creativity
Intuition
Logic

It is your frontal lobes that produces/generates/calculates
CICIL.

When your frontal lobes makes CICIL, your amygdala gets tickled.
Your amygdala is tickled when it sees you are going in the direction of
SURVIVAL- which automatically happens when you are using enough

Cooperation-Imagination-Creativity-Intuition-Logic

When your amygdala is tickled, this allows more energy to easily flow into your Big Smart Frontal Lobes.

When your frontal lobes are happily at work, that tickles your amygdala even more.

It's a HAPPY CIRCLE OF BRAIN ENERGY

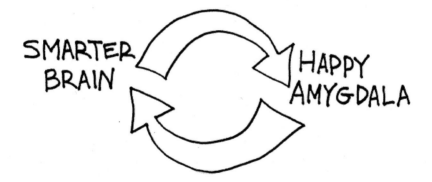

SMARTER BRAIN

HAPPY AMYGDALA

Frontal Lobes **COOPERATION**
Tickles Your Amygdala:

Togetherness is more fun, smarter and better than being alone.

COOPERATION with THE BIG PICTURE
Tickles Your Amygdala:

Partner Family Planet
Team Power for sustained long term survival.

Frontal Lobes **IMAGINATION**
Tickles Your Amygdala:

Imagine a FEATHER tickling your amygdala
ANY TIME
ANY PLACE
ANY WAY

Hee hee hee!
Ha ha ha!

IMAGINATION
Frees you from the LIMITATION of your SENSES
so you can see OUTSIDE THE BOX in a 360 degree view.

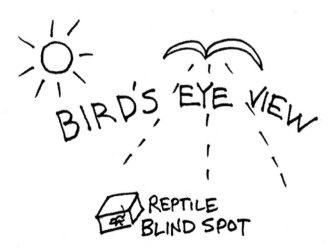

Frontal Lobes **IMAGINATION**
Tickles Your Amygdala:

You see the effects of TIME.

IMAGINATION
Lets you plan ahead

Frontal Lobes **CREATIVITY**
Tickles Your Amygdala:

You can see things in novel, fresh, new, and different ways than before.

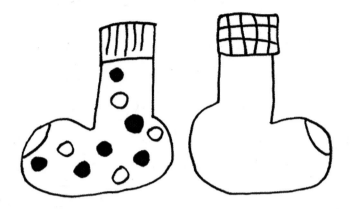

CREATIVITY
Solves problems in unique NEW WAYS.
You combine things as never before.

(Like creating a chocolate covered sledge hammer that you can eat
after you're done whacking nails with it.)

Frontal Lobes **INTUITION**
Tickles Your Amygdala:

It turns on

BRAIN RADAR

You arrive at
The Right Time
At the Right Place
With The Right Solution

Just like Magic.

Frontal Lobes **INTUITION**
Tickles Your Amygdala

Beyond Words - Beyond Time - Beyond Space

Puts you Right On TARGET

INTUITION
Understanding Beyond Words

Frontal Lobes **LOGIC**
Tickles Your Amygdala:

You can focus on how things fit together one piece at a time

LOGIC
Lets you add things up in a linear way, in a row

$$1 + 1 + 1 = 3$$

CAUSE and EFFECT
This follows that.

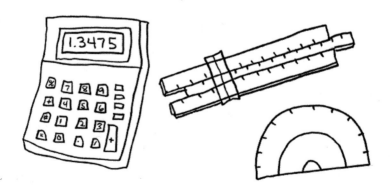

So- you can always use any part or combination of

Frontal Lobes C.I.C.I.L.

Cooperation-Imagination-Creativity-Intuition-Logic

to
TICKLE YOUR AMYGDALA

and always find your
TRUE NORTH

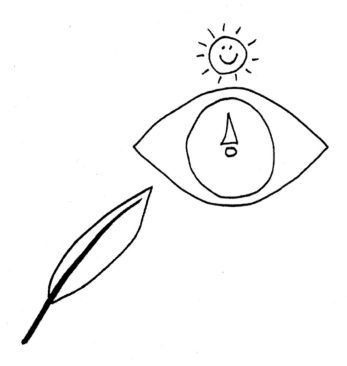

You can
TICKLE YOUR AMYGDALA
by using
the smartest and biggest part of your brain,
Your Frontal Lobes

to
get
PERPETUAL
L.I.F.E.
REWARD

To move A-HEAD
in the RIGHT DIRECTION

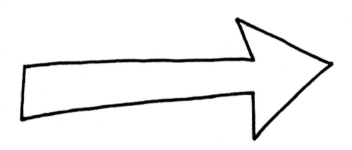

HOW WILL
YOUR
FRONTAL LOBES C.I.C.I.L.
TICKLE
YOUR AMYGDALA
TODAY?

Chapter 5
TWO EYES IN YOUR BRAIN

THE LITTLE ME EYE

You have two "eyes" inside your brain, just like you have two eyes on your face. But unlike your two physical eyes, your brain's internal eyes see the universe in two very different ways.

One eye sees the universe as if you were the center of everything. This is your "Little Me Eye".

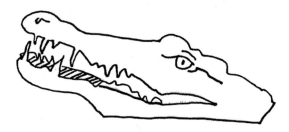

Your Little Me Eye is what you just see with your senses, and what is more or less sitting one inch in front of your snout, give or take a few yards. The Little Me Eye is only concerned with how things outside you affect You alone.

It is the reptile brain that sees through The Little Me Eye the entire universe moving in a one way street towards its ever hungry jaws.

The reptile brain perceives everything with the Little Me Eye in one direction only, a tiny slice of the big universe, with nearsighted vision.

This is good when your stomach is growling, "Time To Eat", but not much more than that.

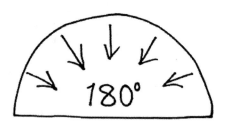

THE BIG MAGIC I

The other "eye" inside your brain sees the universe as you are connected with everything.

This is your "Big Magic I"

Your Frontal Lobes Big Magic I sees how everything, every molecule, every atom, every sub-atomic thingy-a-ma-jiggy is connected to everything else.

The Big Magic I understands how your Little Brain is connected to all the other brains beyond your own individual self. Your Big Magic I extends beyond the boundaries of your own personal body, space, and experience. Your Big Magic I expands 'You' infinitely as far as you dare to think.

The Big Magic I sees in ALL directions, in front, behind, to the sides, into the future, into the past, and on both sides of a coin.

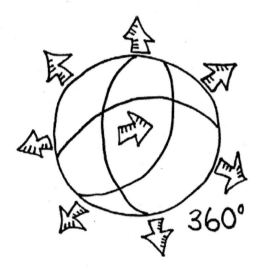

For the longest time people believed that the entire universe revolved around the Earth, that the Sun and all the stars in the heavens all circled us alone on the dinner table of creation.

At one point, Copernicus and Galileo offered the opposing opinion that this wasn't true, and that the Earth and the planets actually revolved around the sun, in a much bigger and grander scale.

At first, this made people laugh and say- "Oh that Galileo! What a dope!"

Eventually the big brains convinced the little brains about the big picture.

If you look at your amygdala with just your Little Brain, your survival will eventually be threatened. You will eventually end up with negative emotion.

That's because you're only using a small portion of your brain potential: The universe through the tiny window of the Little Me Eye.

That's a bummer.

That's a bad trip.

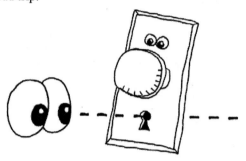

When you tickle your amygdala, you can turn and face the direction of happiness and survival, no matter which direction you start out facing.

To see in all directions- look with your Big Magic I.

What do you see with your reptile brain and its Little Me Eye when your amygdala is looking down the Feel Bad Dead End?

My job sucks.
I don't have enough money to be happy.
I don't like my mate- it's a bad match.
My friends have betrayed me.
I have conflicts with my neighbors.
I had an accident.
I'm sick.
It's wrong.
It's broke.

But what can you see with your Frontal Lobes Big Magic Eye as your amygdala is gazing up the Feel Good High Way?

My job is important.
I have enough money to be happy.
I love my mate- it's a good match.
My friends do what is good for me.
My neighbors compliment me.
Every accident has a silver lining.
My body is doing what it needs to do.
It's alright.
It's working as it should.

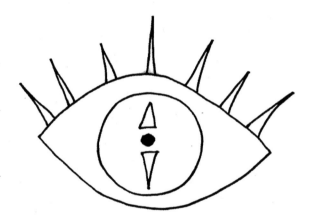

What "eye" are you using to see your You-niverse?

PART TWO:

A Delicious Three Course Amygdala Tickling Gourmet Presentation

1) Details of Amygdala Tickling Brain Radar, Etc.

2) Amygdala Interviews

3) Ways to Do It

Chapter 6
THE AMYGDALA

You are a complex living organism. Unlike an amoeba that only lives about as long as a couple of old episodes of "I Love Lucy" or the time it takes you to drive to the dry cleaners and back, you stick around for a relatively longer period of time, plenty of time to get into lots of trouble.

The complexity of you started long before the advent of even crummy looking black and white TV programs, something like sixty-five million years ago. This was when Mama Nature started equipping Earth mammals with a brain that could flip a whole lot more kinds of tricks than what a peanut-sized dinosaur's brain could accomplish, such as merely biting the head off of the smaller dinosaur who lived next door.

We know what happened to the dinosaurs, and it wasn't so good.

The mammals, on the other hand, thrived and survived to this day, to make pests of themselves, not only by invading the area behind your kitchen stove but also by running for government jobs every couple of years.

One of the reasons for the long success of mammals is a little bit inside the mammal brain that allows furry creatures to experience *emotions*. Obviously, this emotional brain area is not always functional, as observed in those upright humanoid brains that would cut off funding for badly needed social service programs without blinking an eyelash. But otherwise, this part of the brain is a working little gold nugget of neurophysiology that provides a primary function in the service of species survival.

This emotional hub of this furry mammal brain is called:

THE AMYGDALA

Your dog has an amygdala, a ferret has one, a mouse has one, and so does an anteater. For that matter, you do too.

The amygdala gets its name, meaning "nut" in Greek, because it looks something like an almond or a walnut, not because it looks like Jerry Lewis.

You've got two of them, one in each side of your head, one for each hemisphere of your brain.

The amygdala is a hub for forming, retrieving, and processing emotions.

The amygdala is connected to major other parts of the brain. This includes the core reactive parts of the brain which regulates basic body functions as well as the most advanced parts of the brain responsible for complex thought production, abstract creative thought, and social behaviors.

The interaction of all of these areas of the brain results in your emotional response to sights, sounds, sensations, as well as more complex ideas and cues.

EMOTIONS

To a large extent, we make our decisions largely dependent upon how we feel emotionally about one thing or another.

Emotions are a neural shortcut that helps you to quickly discern what is good for you and what is bad for you before you stick your finger in a hot flame or before you bite into that yummy looking cupcake.

Your amygdala lets you quickly react to danger and avoid it. At the same time your amygdala and your emotions can quickly identify what is good for you and you can embrace such things.

In this sense, your amygdala and your emotions can be seen as a rudimentary brain compass that employs "Emotional Magnetics".

At the most fundamental level, you are attracted to those things that have you feel positive emotions and you are repelled by those things that have you feel negative emotions.

You have a magnetic compass in the middle of your brain. Son of a gun.

AMYGDALA TICKLE #1- "Brain Magic Compass"
Go to a toy store and buy a cheap compass. Carry it around in your pocket.

Also, every time you touch it or pull it out and look at it, it will remind you that you have a compass inside your brain.

When you look at it, you'll think about the compass you have inside your brain- and this will begin to tickle your amygdala forward.

EMOTIONAL SHORTCUTS

As a general rule, you are attracted to things that are good for your survival by positive emotions, and you are repelled from things that are bad for your survival by negative emotions.

By no coincidence, the amygdala is actually directly connected to your olfactory nerves, and your sense of smell. You can easily observe how odors can trigger emotions simply by taking a whiff of something you enjoy, and comparing how you feel when you smell something you do not enjoy.

That's your amygdala- telling you to run towards fresh buttered popcorn wafting out of the bowl sitting on your kitchen counter and telling you to run away from the moldy yogurt that is crawling out of your refrigerator next to it.

In the same way that a mammal can smell danger or food far in advance of being face to face with it, your amygdala serves as an "early warning system" so that you can respond long before you understand something.

Your amygdala in this sense helps you to "sniff out" threats and rewards, even when you haven't yet figured out why you might like or loathe something.

This emotional evaluation that you make of things that cross your path happens instantly through your amygdala, faster than you can rationally think about such things. That's why you still have an amygdala and why it's never been discarded into the evolutionary garbage heap.

AMYGDALA TICKLE #2- "The Nose Knows"

To understand the difference between clicking backwards and tickling your amygdala forward,

FIRST: Find something in your house that smells bad or unpleasant. Stick your nose in it. Congratulations. You just experienced an Amygdala BITE.

Now, TICKLE YOUR AMYGDALA- Find something that you like the smell of- popcorn, a flower, your favorite food, your dog. Congratulations! You just tickled your amygdala forward!

But your amygdala is connected to all of your senses. So even though you can't smell it, when you see and hear an asteroid flaming down on you through the clouds, you don't think much about it- Your amygdala instantly tells you, "This is not good. Run like heck!"

DETAILS: TICKLE YOUR AMYGDALA

To "Tickle Your Amygdala" means to consciously and directly stimulate the reward pathways and to activate the pleasure circuits of the human brain, not from random or external causes, but by personal will. You tickle your own amygdala, rather than have it stimulated by accident by whatever you happen to run across.

This is done by consciously invoking the brain's basic frontal lobes processes of

Cooperation
Imagination
Creativity
Intuition
Logic

These C.I.C.I.L. processes occur singly or in any combination determined by the individual's preferences and needs.

Your frontal lobes create a positive force expressed by CICIL: Cooperation, Imagination, Creativity, Intuition, Logic.

Cooperation: Constructive Social Interaction
Imagination: Abstract Thought, Ideation of Time
Creativity: Syntheses, New Combination of Disparate Elements
Intuition: Non-verbal Perception, Non-Linear Ideation
Logic: Linear Perception, Verbal Ideation

Apply one's frontal lobes skills completely, and you get the whole kit and caboodle- no reality hangover later on.

Of course, one can apply one's frontal lobes skills in an incomplete manner to continue to allow or to even cause pain and suffering, either deliberately or from incomplete awareness and calculation that will result in an amygdala "Bite" for one's self or for another person.

For example, you can cooperate with Genghis Khan and elicit a negative effect on your next door neighbor.

Conversely, you can cooperate with Mr. Rogers, and this will result in a chocolate cream pie on your neighbor's porch.

The same holds true for any of the CICIL elements.

I.e. you can imagine World War III, or you can imagine Erfie and Chloe's Birthday Party.

You can create a 500 foot tall smelly and thorny lizard by the name of Irving, or you can create a flying bunny rabbit with presents named Gertrude.

You can intuit good luck, or you can intuit bad vibrations.

You can logically formulate a deadly Andromeda Strain, or you can formulate a Positive Emotional Equation such as $E+ = FIQ+$ (Positive Emotion is equal to Frontal Lobes Intelligence).

The difference is, when you just see a little piece of the pie you're only solving a little piece of the puzzle. You're only using an insufficient fraction of your Frontal Lobes Power.

But when the frontal lobes are fully engaged, seeing the Big Picture over the long term and the effects on All- you get a good feeling inside your brain.

If the positive emotion is temporary or leads to a negative rebound and reality hangover later, it is not going to do you much good in the end. So your challenge as a human is to figure out how to keep amygdala tickling going for more than a brief moment. And that's what your frontal lobes are for.

Your frontal lobes can allow you to see around the corner and farther than what is one inch in front of your nose.

In order to insure long term positive emotion the organism MUST employ a reliable system, and that means frontal lobes thinking.

By using your frontal lobes, you can tickle your amygdala and keep on truckin'.

Every person that tickles their amygdala- by whatever method- does it without exception by employing one or more frontal lobes processes of C.I.C.I.L.

Cooperation – Imagination – Creativity –Intuition – Logic

By definition, tickling your amygdala indicates enhanced long-term survival, indicated by sustainable positive emotion. The pleasurable reward response is not reversed by a "reality hangover" and negative rebound after-effects.

Amygdala tickling produces effects superior to temporary and randomly stimulated reward responses that do not involve sufficient frontal lobes processes, and that are subject to negative rebound.

Tickling your amygdala results in superior problem solving abilities beyond those accessed by random discovery, negative reinforcement, or poor methods of learning.

The amygdala tickling reward process positive emotional effect may be learned and repeated to produce a positive habit which is self-reinforcing. The results of amygdala tickling are progressive and accumulative over time.

Josh Blue
Stand Up Comedian, www.joshblue.com

Josh Blue tickles his audiences' amygdala every time he goes on stage. He became a household name in America as the winner of the hit NBC television show Last Comic Standing, (2006). A member of the U.S. Paralympic Soccer Team, he competed in Athens at the Paralympic Games, the world's second largest sporting event.

He continues to tour full time performing live shows on stage and on television, and has several comedy CDs, a DVD, and an upcoming book to his credit. His life and story illustrates many frontal lobes amygdala tickling principles.

NS: "What was a pivotal point in your life?"

JB: "When I was fifteen, my parents and I moved to Senegal, Africa. It was an eye opening experience. I have cerebral palsy, and until then I was very down about it. I couldn't do what everyone else does. But to then go to a third world country and see what people have to do just to survive- I thought, 'Holy sheet, I've got so many good things going for me, my disability is nothing. At least I've got food, and shoes on my feet.' Fifteen is such an influential age, and I couldn't have had a better experience."

NS: "Where does your humor come from? What is funny, and why?"

JB: "People have such a skewed idea of what disability is and what they think I should be capable of. The fact is, I'm usually smarter than the person who is condescending towards me. That's funny to me. It's the bait and switch- 'Okay, you think I'm that person, so I'll just lead you down that path to let you think I'm that person, and then I'll switch it up at the end at a crucial point.'" (laughs)

NS: "What about your parents?"

JB: "My parents home-schooled me. My mom is a librarian, my dad is a language professor, a genius who speaks thirteen languages. My dad told me, 'You don't need to know everything; you just need to know where to look it up.'"

NS: "He married the right woman- it was a match made in heaven."

JB: "My family is very smart, everybody in the family is a teacher, and everybody speaks different languages fluently. All my siblings went to a fancy private school, but because of my handicap my parents had to send me to a public school. It was an experiment, 'Let's send one to a public school and see what happens.' I'm not book smart like my siblings, but I'm street smart and I can beat the hell out of them!" (laughs)

NS: "Well, you make your living out of language, just like everybody else in your family."

JB: "Yep. And I speak three languages anyway, English, French, and Wolof, the native language of Senegal. If we were speaking Wolof, we would still be greeting each other for ten minutes, 'How's your dog? How's your house? How's your family? How ya' doing? You okay?' Just on and on. It's a very playful language."

NS: "Was your humor a mechanism for survival for being different?"

JB: "Humor is the best defense- to make people laugh. If you're going to make fun of me, but I've already said something funnier than anything you'll ever say, you'll look really dumb by comparison.

Although I didn't really fit the norm in school, I could cross all the cliques. I could sit at any lunch table I wanted to. I just chose the all-black girl lunch table, 'cause it was the most fun. Besides that, if anyone picked on me, those girls could just shred anybody, verbally or any way. (laughs)

After college I went back to Senegal for a while and did some independent study as a zookeeper, which was totally random, but life changing. One time I got the zoo to lock me into a cage for an entire day and put me on display."

NS: (laughs) "Okay, so what was that like, what happened?"

JB: "I had no shirt on, just shorts. The thing was, if you take cerebral palsy and take it out of context and put it in a cage, people don't know what the heck is going on. During the afternoon, there would be like seventy people constantly around my cage feeding me peanuts and fruit.

Also, I shared a wall of bars with a four-hundred pound gorilla."

NS: "What did the gorilla think?"

JB: "Well I established a relationship with him. I was one of three people in the world who could pet him. He was a beast, but cool. So I was on the other half of his enclosure. So during the day's siesta, everyone has left the zoo, and I decided to stay in there and do the whole day, just like a real animal.

So I'm half asleep on the concrete floor, and nobody is around and all of a sudden I hear this noise. The gorilla has got his hands around the metal door and he's bending it to get into my half of the cage.

So I had to really quickly start playing with him, running back and forth to keep him distracted so he would stop trying to break into my area, because there was nobody around to save me.

The zookeepers started calling me the 'Boo Boo Monkey' that day, and they were telling people that they had captured me in the mountains of the Congo. And I didn't talk all day. But people were talking about me in Wolof, and nobody thought I could understand them- but it took everything to keep from laughing.

One lady starting talking to me in English and said, 'You're crazy!' and then I suddenly said back to her in Wolof, 'You're crazier than me!' which she didn't understand but everybody else standing around did, and everybody started laughing, and she didn't get it at all.

One time I got malaria, and I had to go to the doctor to get a shot. So while I'm talking to the doctor I tell him, 'You ought to bring your kids to the zoo sometime, because I work there.' So of course, the day he decides to bring his kids to the zoo is the day I'm in the cage next to the gorilla in my shorts. He's probably thinking, 'Oh gosh, the shot didn't work.'

And he's looking at me and says through the bars, 'Do you remember me?' And I'm just in there grunting like an ape going 'Oo oo oo', and I made some ape sign language symbol with my hands for an injection in the butt."

NS: (laughs) "How old were you then?"

JB: "I was about twenty. To be honest with you, at that point in my life I went, 'Okay, now I can die. If I die now, I can say that I've done and accomplished something that probably nobody else has ever done.'"

NS: "For you it was that point where you said, 'From this point, anything else goes."

JB: "Yeah."

NS: "What led you to stage comedy?"

JB: "Somewhere in there I tried stand-up in college. My friends got me to an open stage, and I was so nervous. But a cool phenomenon happened. I went up and did ten minutes of story telling to about twenty people. I said I'd come back in a week, and through word of mouth seventy people showed up the next time just to see me.

The last quarter in college I set up a curriculum to actually study standup. I set these goals of what I was going to do. I studied Richard Pryor…

There's a difference between when you watch comedy just to enjoy it, and when you're thinking, 'How did he stand when he delivered that joke? How did he set that up?"

NS: "Left brain, very analytical…"

JB: "Right. 'How does he go from the set up to the punch line?' So, from there I decided to do a show a week, and from there hopefully go on stage to a real comedy club.

So what happened was I found this band that was performing at this coffee shop. The place would have two bands. So the headlining band said I could do the fifteen minutes between the opening band and the headlining band. Very rapidly everybody realized that the coffee shop would be full for the first band, and then me. And then everybody would leave after I was done.

I felt really bad, because the band that invited me was left with like two people in the audience.

So the coffee shop owner said to me, 'You need your own night.' And so they gave me my own show, and I did an hour show every week of new material. But now that I'm a professional, and I can't do that any more!"

NS: (laughs) "Why is that?"

JB: "I don't know, I've never written anything down."

NS: "I've had nightmares about that, being on stage, forgetting everything."

JB: "Here's my theory- If you don't write it down, you can't mess it up. If there's nothing written down, how can you mess it up? There's nothing there to mess up.

I would just tell stories, I'd tell stories from Africa, stories from the zoo. But I never practice, I just think about what I want to talk about. I just show up. It's just a gift I have. I can talk about anything and make it funny."

NS: "That's extraordinary. I'm a musician- that would be like me showing up to a gig without having any music, and just play for an hour improvising. I guess Cecil Taylor used to do that, but he sounded horrible- my opinion."

JB: "I just look at it as the story of my life that I'm telling about. I laugh when I look at things, it just comes out on the spot."

NS: "Tell me a bit about *Last Comic Standing*…"

JB: "I just moved to Denver on a whim. I got a job as a counselor for disabled people, but I was miserable at it. I knew there was more to my life than this."

NS: "That's your amygdala signaling to you through your emotional feedback system, your emotional pain that there's another path that's better for you."

JB: "Yeah... 'This is not me.' So I just kept saving my money, and saving my money.

There was this one co-worker who used to get on my case because I'd pick up money in the street when I'd see it. So he'd come in and bring in a pocket full of pennies and he'd just throw 'em on the floor and watch me pick them up, I'd say, 'Hey this is money, I don't care, this is enough for a beer tonight!'

But I'd say, 'I have to get out of this...' And we would come up with these evil schemes. I have this charming side, and I can talk anybody into anything, but in the back of my mind I'd hatch these evil plans.

But then I discovered the comedy scene in Denver. I was looking at the marquee outside the Comedy Works, and this couple comes by and I pointed to the sign and said to them, 'One day my face is going to be up there on that marquee.' They looked at me and said, 'We believe you!' It was weird, but I could tell, they really believed it.

So I went home and said to myself, 'Let's figure this out.'

I just started going out to these different clubs, and I ended up working my way up at Comedy Works from two minutes, to five minutes, and then I finally got promoted to a paid position there. I won a couple contests, and then I quit my other job.

I finally got a manager after word got around, and he started booking me on the road. Finally I made 'The List' at Comedy Works, and I was opening for all these big comedy superstars. For me, it was like school, I would come in five nights a week and study the headliners. What an opportunity for a young comic to go and do that! What I learned was what I like about comedy, and what I don't like about it.

I saw how a headliner conducted himself. Some other comics are such arrogant sons-of-a-beech-tree, they think they're special cases, but it works against them. But when I saw Brian Regan, whenever somebody came in the room before the show, even if it's just the door guy, he would stand up and shake his hand, 'Nice to meet you.'"

NS: "He's interacting like any normal person might, making an effort to be personable"

JB: "Right. And I learned that the waitresses at the Comedy Works have a lot of power. A lot of young comics don't realize that. If you stay out of their way and tip well, they go to bat for you with the audience. I mean, obviously you have to have some talent too...

Being funny is not the whole thing. There are so many other factors to the whole machine. That's what I was learning, how to conduct myself off stage.

To me, interacting on a nice friendly way is just how I conduct myself, but I realized on a professional level that it's beneficial."

NS: "To compare it with lions, hunting in a pride- One lion goes out by himself, he's not nearly as successful as if he's got all his friends with him. And everyone has their own role to play in the hunt, but it's this cooperative mind-set, and it proves more successful."

JB: "Last Comic Standing went on from there. I auditioned twice. First year I was in Chicago, I waited in line for six hours. I remember they had one of those clock-thermometers on the building, and it was a negative fifteen degrees."

NS: "That's insane."

JB: "You're jumping around trying to stay alive outside in the line. I barely thawed out by the time it was my turn to go on. There are these two guys in the audience and five cameras. Comedy is a very give and take kind of thing. Even if the two guys are laughing their hardest... I only did like a minute and they said, 'Okay, thanks... next.'

So as I'm walking out I actually said something that made them laugh, 'This is a shitty make a wish.' And they laughed at that.

The next year came around, and I was very jaded towards the show because it wasn't really serious a tryout. But my manager managed to get me a real audition where I didn't have to stand around in line.

But I'm glad I didn't get picked the first year, because in between the first and second, and did like thirty college shows and I was that much better.

Once I got on there I just went on with the attitude, 'This is mine to win.' And I did."

NS: "How long was the process, from the first show to the final win?"

JB: "I was on for thirteen episodes. And the thing was that they extended it because we had such good ratings."

NS: "How did your life change because of all of this?"

JB: "I was very fortunate because my management had the attitude that I was going to win, so we prepared for that. We talked to the guy who won the year before, and asked him 'What do we need to know?' So it was a smooth transition when it happened, we had everything in place."

NS: "This is something that comes up a lot when I talk to people that are successful- they say, 'I act as though I already have it. I'm in that place.' And that changes the way you operate from where ever you are, before you actually get it, and is much more conducive to succeeding."

JB: "I have to say one other crazy thing that was going on during all of that, I was trying out for the national Paralympic Soccer Team, and I got on the team and went to eight countries as a professional athlete, and our team just qualified for the London in 2012. Athletes dream of what I've accomplished, but for me that's just the backburner to my comedy career. That's just insane. I mean, I'm thinking, 'How do I stay in shape when I'm on tour and drinking beer each night?'"

Whatever it is in my head, it's just that I will not be deterred, I will not be denied. I'm not in the best shape on the soccer field, but I know how to kick a ball and know how to get it where it needs to go."

NS: "Have you ever given much thought to the specific things and processes going on inside your head that allows you to do what you do?"

JB: "There's a couple things- Because of my disability I've had to learn alternative routes to get to an end product that aren't the normal way of getting there, like in writing a book, using someone else to dictate. I think that's one reason I have success at what I do, because I don't see things the same way that everybody else does.

As far as feedback, I feel like I'm now just embracing the effect that I have on people. When you see my show you're going to get a hard laugh- but you're also going to leave with a different perspective of a disability and the disability community.

I'm realizing that there is more to what I'm doing than just comedy. At least once a show somebody comes up to me and says something like, 'My dad died of cancer last year, but he watched you and we had a laugh together.' That's real, and that's heavy.

My personal opinion is that everybody has some sort of disability of some sort or another, and so everyone can relate to what I'm doing."

AMYGDALA TICKLE #3- "Laughing Lobes"
Humor tickles your amygdala- because it requires frontal lobes thinking- anticipation and imagination, and creates positive emotion. It's good for you brain and for the many positive effects on your body and mind.

Get yourself a funny book, go see a funny movie, learn and tell a funny joke. Keep yourself laughing. This will tickle your amygdala in more ways than one.

Chapter 7
POPPING YOUR FRONTAL LOBES

Tickling your amygdala results in superior problem solving abilities.

It culminates in the "Eureka" phenomenon, a positive peak emotional experience in which an individual suddenly becomes conscious of a hitherto unrecognized or unformulated solution to an outstanding and usually difficult unsolved problem.

This Eureka moment is called "Popping Your Frontal Lobes" since it is the result of applied aspects of recognized frontal lobes processes of the human brain. It's "popping", maybe in the same way you experience a great relief when your ears finally pop at high altitude and all that internal pressure is so delightfully released. Whew. So happy

Popping Your Frontal Lobes is an experience that clearly stands apart from the more or less ordinary collection of daily experiences and emotions.

Notably, Popping the Frontal Lobes can occur in individuals who are not familiar with the terms "Tickling Your Amygdala", or "Brain Radar".

Importantly however, those who *do* have an understanding of these terms and the mechanisms of these processes can consciously invoke these processes, and thus cause popping of the frontal lobes to occur as an inevitable occurrence in their own experience as opposed to a completely chance phenomenon like stumbling into a $20 bill lying on the sidewalk once every forty-seven years.

Popping Your Frontal Lobes can occur any given number of times in one's life. The experience can vary from individual to individual as well as also vary in intensity.

Popping Your Frontal Lobes can be a physical sensation as well as an emotional or intellectual experience, or any combination of the three elements. It is usually marked by an overall feeling of contentment, elimination of anxiety and negative emotions, as well as a palpable general feeling of euphoria and sense of well being.

"Gee whiz good golly Miss Molly, dot feels reeeeeeeeaal nice!"

The moment of Popping Your Frontal Lobes is *specifically* unpredictable. However, it can be predicted *in general* to be an inevitable occurrence when an individual can readily identify and deliberately invokes frontal lobes processes.

Robert Schneider
Professional writer, www.writing-resources.org, "A Cookbook of Consciousness" (Sihanoukville, Cambodia)

NS: "You've written some extensive articles about the amygdala…"

RS: "Um, I've been reading about clicking and tickling the amygdala since 2003 when I first wrote to you about it. I had found your site through a rather circuitous route, first reading about the amygdala on one web site that pointed to another, and eventually to yours. I had been living in Australia since 1985 after moving from the US, and I hadn't heard of you previously."

NS: "How did you end up in Cambodia? Slow boat from China? Heh heh… [Robert has lived in Cambodia since 2006.]

RS: "I had always wanted to live in a small country. I went looking, and ended up here. I'm now 63, and I write content for various web sites."

NS: "Since you've written a lot about tickling the amygdala and because you've been doing it a long time, you're a good person to ask- What does amygdala tickling do for you, and as Gerry and The Pacemakers also asked, 'How do you do it?'"

RS: "I anticipated this question because I knew we would be talking. I'll go back to the beginning, the first time I tried it-

When I was in my early twenties I had studied yoga, and I had worked as a yoga teacher and as a meditation leader at a retreat. When I reached my thirties, I was over it…"

NS: "Hold on. What do you mean, you were 'over it'?"

RS: "First, I became disenchanted with my supposed guru, and then I became disenchanted with the man who ran the retreat. Basically, I didn't have a life because I was meditating so much, and that got in the way. Then I moved to San Francisco and started working. Essentially I wanted to have a life outside the yoga community.

But also everything changed during the late '70s and early '80s. By the '80s hatha yoga turned into a way to get fit and meditation turned into a business. Werner Erhard and EST came along and everything turned into a business."

NS: "It all turned into 'me me me'…"

RS: "Yes, and I wanted to start a family and I had to think about practical things. Then in the late '90s I started getting interested again."

NS: "Maybe you were looking for a middle ground."

RS: "That would be it. I was exploring again.

When I had been meditating I had a few extraordinary experiences that made me sure that things like Kundalini and the chakras were real. I had experienced psychic phenomenon, but I wasn't looking to make any of this stuff a lifestyle- I was looking for a middle ground, I guess.

I had been looking online for year and had learned a few interesting things before I stumbled on your web site. I read about this amygdala and I was impressed that you shared this simple technique for free, and…"

NS: "Well, I sure ain't a millionaire from it!" (laughs)

RS: (laughs) "I was a little dubious about the science at that time, because it was so easy to do the technique. But I decided to try it because I just happened to be at the absolutely lowest time of my working life.

I had my own business until 1994 making custom furniture, and then I had to give that up because of a bad back injury. About the time I had stumbled across your site I had lost a decent paying job just overnight, and I had to go work in this horrible, horrible factory job where they made these multi-million dollar yachts- which sounds interesting, but it was just an absolute hole.

98

That was the first time in my life that I was chronically depressed, not clinically, but chronically, like I had screwed up my whole life.

It was right at that time that I had read about amygdala clicking, and had been trying it now and then. It made me feel mildly better. Even the first few times that I did it, I could definitely feel a slight release from pressure. But nothing extraordinary.

But one day I was driving to work, and this was at about 5:30 in the morning, freezing cold. I was looking forward to going into this even colder, massive factory where the smell of chemicals was everywhere and the air was thick with fiberglass dust. And the crew did charming things like, when they'd get in a fight one guy would staple another guy's hand to deck of a boat." (laughs)

NS: (laughs) "Good god!"

RS: "In Australia, if you were on welfare and on the dole you have to look for a job and report. If you can find a job you have to take it. This place was *so bad* that the welfare agents would say, 'No, you don't have to take *that* job.'"

NS: (laughs) "I'm sorry- but it's the way you tell it, Robert..."

RS: "It was the only job that I could get in an instant, and I was broke and had a mortgage and everything. It was a terrible time.

But anyway, one day I was driving along, thoroughly depressed, and I did a little amygdala click and became completely blissed out.

The freezing cold air felt invigorating. And when I got to the horrible workshop I thought, 'Wow, this place is fascinating. It's really interesting."

NS: (laughs) "You weren't taking any drugs, correct?"

RS: "No. No drugs whatsoever, but it was as if I had taken a very strong one- it was that big a change. That feeling persisted for a good six or eight months I guess."

NS: "You're talking about a permanent change?"

RS: "A permanent high. Yeah. Every time I clicked forward I'd get on a big high.

That job ended after three months, and went to a similar job under much better conditions.

Yes, it was just quite extraordinary. Then it tapered off, and things got fairly normal. But after that, things in my personal life went from bad to worse."

NS: "In spite of the clicking?"

RS: "In terms of my *outside* personal life, things went from bad to worse. My dad was dying, so I had to quickly go to America. I had to stay there and take care of him for three months until he passed away."

NS: "I have to ask you this- You were able to click forward, which improved your *internal* emotional state, correct? This was in spite of the fact of these external circumstances that became more and more difficult."

RS: "Yes."

NS: "So then, what you're saying is that it was two different things- You were not affected by the disintegration of what most people would consider 'a good life' or good relationships, or whatever it was. But you were still able to maintain a good equilibrium and a positive outlook even though outwardly things were falling apart. Is that accurate?"

RS: "Yes, you put it exactly right. I was able to maintain equilibrium and a positive frame of mind, even though my whole life was falling apart around me.

On top of everything else, when I came back to Australia, the first thing my wife announced was that she wanted a divorce after twenty years, and she moved out of the house. I'm not saying this to disparage her, it's just what happened. I was absolutely stunned, and there were some times that I was freaked out- but it never overwhelmed me. It was just, 'Okay, I've got to deal with it.'

So she moved out, and my daughter was finishing high school, so I dealt with all of that.

I know for a fact had all this happened and I had not learned this absolutely ridiculously simple technique, I certainly wouldn't have been able to handle things the way I did."

NS: "People think that the good feelings that we have inside our brain are dependent upon external circumstances. That's the tendency- which I think is a cultural brainwashing."

RS: "Yeah."

NS: "Obviously, if you're doing a job you hate you're not going to be happy. But previously you actually had a job that you hated, or thought you hated, and you tickled this switch in your brain and suddenly you perceived that job in an entirely different light so that you were able to get something positive out of it."

RS: "That's exactly right."

NS: "People think that if, 'I've got the perfect mate,' and 'I've got a new car that runs perfectly,' and 'I live in a big house,' and blah de blah de blah, *then* I'll be happy."

RS: "But that's an illusion."

NS: "Your story proves this isn't the case in a remarkable way."

RS: "It is actually. And I can say that without bragging. It was simply that flip, that simple little flip of the amygdala. This is what is so extraordinary to me, that it happened, and that there wasn't anything subtle about it. It was just a complete change of outlook.

Since that time, I haven't experienced such a dramatic change so suddenly, but I can see that it continues to happen even to this day. When I get a little bit down at work, and then just flicking the switch, and flip forward to make that little leap and another point of perspective appears.

It helps you get out of a rut, and lets you see things in a little different way. There's this internal mechanism there that will help you to do the right thing for you at that moment in time.

It just seems like it's just a mechanism, that's why I think it's such an extraordinary technique. It doesn't depend on this long involved self-analysis. I mean, you can do that too- but this doesn't necessarily involve this long analysis, or sitting down to meditate for an hour, or taking an anti-depressant or something. It's just tickling a switch."

NS: "What I know after so many years and so many stories is that everyone's experience is different. At times people will have the kind of experience you're relating to me, perhaps a very dramatic event. And at other times, people may have just a very subtle effect. But essentially, tickling your amygdala allows anyone to instantaneously step back and stop a reptilian reactive brain process that is so ingrained either through habit or cultural conditioning.

It gives a person the ability to tap into the infinite capacities of the frontal lobes. You can turn off the fight or flight, or ..."

RS: "Or at least put it on hold... so you can get a breath."

NS: "...Put it on hold, so instead of being controlled you begin to control. At least its one very simple tool that anyone with a brain has access to at any given time.

So, when you flip forward, what is the process for you?"

RS: "Images are the only thing that works for me. I used to have terrible visualization skills. My mind was all verbal and left brain, and I couldn't hold an image in my mind. But that's one thing that just changed dramatically since I learned this technique."

NS: "What do you see when you do it?"

RS: "It changes. Right now I'm using the feathers, like I did in the beginning. I started by using one feather, because I couldn't hold two in my mind. I also had to do something tactile when I started. When I started to do it, I placed my fingers on my head so I could feel the pressure. Then I imagined one feather- first I gently tickled my left amygdala, then the right. At the same time I tried to let go of any tension that I had in my body. So, I had to use that touching, tactile thing at first. But after that; big whammy.

It was a matter of weeks, or a month or two before that big bang amygdala Pop! happened."

NS: "I don't think anybody ought to worry about this 'big' experience. Probably one day when you're not thinking about it, you might just get that big light bulb going off in your head. But you've described to a T- that very thing. So, you weren't expecting it to happen ..."

RS: "Oh, not at all. Driving to work I was bored and depressed. By then, I could visualize that feather. But here it was, it was raining, and I can't describe how dreadful the rain and the freezing cold was at the time. I just randomly tickled with the one feather, and then BANG! just like that.

Since then, I'm able to visualize anything, and I just let whatever come to me."

NS: "Oh, so your visualization skill changed at that point?"

RS: "Yes. Dramatically. Dramatically."

NS: "Now forgive me, don't be insulted- but you've never been under any serious psychiatric care?"

RS: (laughs) "No, no, no."

NS: "...Because I know some people are going to hear this story and think, 'Oh, that guy is a nut!'"

RS: (laughs a lot) "I suppose some will."

NS: "You're a writer, you're a regular guy, you make your living as a writer, you don't belong to any cults."

RS: "I was a cabinet maker…"

NS: "Gee, maybe it was all of those yacht fumes…"

RS: (laughs) "No, not that either. But, if you're writing this down… I have to tell you, other visualizations would come to me- like this little fairy would come, and the tips of her wings would tickle my amygdala…" (laughs)

NS: "Oh, that's not so nuts. Seriously, lately I've been visualizing a butterfly, and I've been telling people about it. 'Look at me, fellas!'"

RS: "Yeah, yeah. Things like that. Soft gentle things. But these images come to me automatically; I don't seek them out now. Anything like that, gentle and light.

But other times, there might be something like two laser beams coming together, but its usually soft gentle images that work the best."

NS: "When I first learned to click my amygdala up at the brain lab, the director suggested imagining two laser needles, and sticking those in my amygdala. And I cringed when I heard that, and went, 'Uh, I don't think so.'

So I thought of this image of tickling my amygdala with a feather. I wanted something that would feel good, like a feather. It's going to feel pleasurable, rather than shunt this switch forward like an electric switch. But I guess sometimes it is like a big bolt of electricity too.

I'm sure there's probably people that would prefer a laser beam or something. This is a do-it-yourself thing, it's not exactly one-size-fits-all. You know? In fact, the other day I talked to Stevie Wonder's piano tech of ten years, and he used a tuning fork to tickle his amygdala."

RS: "Oh wow… That's the beauty of it for me. I wouldn't need that much practice- and then you can do it any time, any place. All you have to do is try it, and you're doing it."

AMYGDALA TICKLE #4- "Basic Feather Tickle"

Imagine you have a feather inside your head.

With the tip of your feather, touch the tip first to your left amygdala and tickle it. It glows warmly and sends energy to your frontal lobes, which also begins to glow with energy.

Then tickle your right amygdala. It also lights up and sends energy to your right frontal lobes.

Try it with one feather, stimulating and tickling both amygdala forward in unison. Whooosh.

Shari Harter
Self described "Enquiring Mind", S.F. California

NS: "You wrote to me and described your Frontal Lobes Pop! experience. Can you describe that again for me?"

SH: "I do remember that all that amygdala stuff made sense to me from the beginning. But anyway, I was in the bathtub one night, and I was in a relaxed self-contained place and I was doing my visualization there, and I came up with my own thing-

I'm comfortable visualizing stuff in the center of my head, so I would see a butterfly in the center of my head, maybe where the pineal gland is, and the wings would flap forward to the front of my brain. Their wings would stick straight out to the side, and then meet at the front, and then go straight out to the side, and the meet at the front again, moving like that.

That's how I would tickle my amygdala. I could stay focused and do that for quite a while, and then I had an inner Pop! And recognized it."

NS: "Um, it wasn't something else popping in the bathtub water... you know what I mean? A little bit of... (raspberry noise)"

SH: "Haha! That's funny. No. It wasn't that."

NS: "How did you get interested in the brain and what different kinds of things you might do with it?"

SH: "I had some physical problems when I was a kid, and so I was bedridden for a year. I couldn't keep up with my friends physically, and I began to read a lot very early. In the first grade I was already reading fifth, sixth, and seventh grade books. I just lived in my head a lot."

NS: "Your physical handicap led you to develop these other parts of your brain in lieu of running around-"

SH: "I would say so. And when I did play, it was a quieter kind of play. For example, my neighbors had an avocado grove. My favorite thing would be to walk around the grove like a Native American, and not make any noise or leave footprints. And what I would do is start to see these spirits, and I would talk to these guides who would explain things to me and show me around."

NS: "Some people would say you just had a very active imagination, and other people would say you were communicating with spirits."

SH: "Sure. My mom was interested in reading about psychology, Freud, Jung, and Adler. Those kinds of people. So consequently that's what I would be reading too. I was reading psychology in elementary school. She would also get these other more far out books like Edgar Cayce, Bridey Murphy, Jess Stern. Things like telepathy and ghost stories.

I was very aware that our culture didn't even acknowledge a lot of these kinds of things, and I began to wonder why it just shoved a lot of these things in a box and buried it. It made no sense at all to me.

I finally realized before I was very old, that for the last five-thousand years we've been living in a dysfunctional culture. I realized that to be sane in a dysfunctional culture, you would look insane to everybody else."

NS: "Now I understand why my neighbor looks at me so funny. So how does brain self-control and amygdala tickling fit into all this and your picture of things?"

SH: "Because I do have such a strong left brain, I have to tear everything apart and analyze it. But I also believe that the universe 'out there' is a projection of what is inside yourself. So I believe that the most important thing is to align with your higher self."

NS: "How do you do that?"

SH: "A lot of that is just relaxing and allowing- I know my left brain is clutching and defining and judging and all that. I just imagine all of that and clear the slate. I imagine the entire universe turning into cosmic foam, and I just take all beliefs and thought forms about how it is, and I just wipe the slate clean. So you first go to a clean slate. I clear out all definitions as much is humanly possible.

I do see a vertical line running through my body from the top of my head going all the way down to the earth. Then I see a vortex in my heart, a glowing holographic flower, like a golden light, and I'd bring the energy from my heart and send it outward. This is like a portal or a wormhole from me to the heart of all-that-isness, the source, the heart of my being connected to the heart of the creator.

I just imagine that it is like a fountain of youth, just bubbling away and bringing all that energy into my life, the heart field expanding and going before me making perfect my way."

NS: "You project that forward in front of you..."

SH: "More like a 360 degree ever expanding radius around me, broadcasting in waves, connecting with everything."

NS: "Okay."

SH: "I have a few little mantras that I've picked up along the way, and I use those too. You know, something as simple as 'Everything is falling together,' because it's so easy to see that everything around me is falling to pieces. You know what I mean?"

NS: "You mean everything *isn't* falling apart? Why are my taxes going up every year then?"

SH: "Because you know, I just let myself know that no matter how it looks on the outside, my higher self knows better what I should be doing, and where I should go and like that. If I end up on a detour, I know that it's what should be happening."

NS: "For example, if Neil Slade is forty-five minutes late in calling you up for a phone interview..."

SH: "Yeah, I usually don't get upset about that kind of thing. The other thing that I'm doing is ceasing to push against things, because that just seems to give power to what it is that's trying to push against you.

That whole idea of confrontation is used by those who would use draconian measures, by the wolves that are the pros at that sort of thing."

NS: "Confrontation can be a double edged sword."

SH: "You can't solve it at the level of the duality. If you engage with them at that level- then they've won. They set it up, they will set you up, and they will win there.

The only place to go is to the level of consciousness above that, to the unifying consciousness. And then you pay no attention to what you don't like, and just go to the place of the clean slate and just start imagining the world that you want."

NS: "Well, if you're clicked backward into fight or flight, into attack-counterattack mentality, even over a little thing, your amygdala has hijacked your brain and you're trapped at that ground level and all of your higher processes of problem solving are dead in the water."

SH: "The work that you're doing is letting people very quickly get out of fear, which is very valuable.

But I do have a left brain which is very active, so I let to get in there and figure things out, and where I'm going, and how I'm going to do it- you know, control, control, control. And somewhere along the line I realized it's there to help us get around in our physical body- but its not there to help us plan our lives, figure out who we're going to fall in love with.

So what I would find myself doing is to tell myself, 'Give that dog a bone, before it chews up your life!' And that's why, for example, I've sometimes seen psychic readers play solitaire while they're doing a reading. They have to give the left side of their brain a game to play, with numbers and logic and get it out of the way, to literally give that part of their brain a job to do so the other part of their brain can tune into the bigger picture."

NS: "Most of what our brain does is non-verbal and non-linear to begin with, there's no question about it. But we also have to recognize that orderly part of our brain has a function.

For example, what if there was no card catalog system? How would you even find your favorite book at the library? If you wanted to bake a cake, and if you didn't have a way to organize your kitchen cabinet- you couldn't find the ingredients that you need when you need them. Your flour would end up in the refrigerator and your sugar inside the oven."

SH: (laughs) "And I have a friend like that I swear! But you have to use each side for what it does, and you use both.

I have a big collection of dance music of all different kinds- and so I just start dancing, not ballroom dancing- but I just start boogie-ing- and you just let your body move the way it wants and you feel your spirit like that.

But, are you familiar with the Morgan's Tarot deck? It was done by some guy in the Santa Cruz mountains, but there's this one card that shows the top of a guy's head, and there's a switch on the top of his head. And the label for that card says 'Go ahead, On- You have the switch." (laughs)

NS: "Oh yeah- and I used that image for years, of an electrical switch in your brain, for the amygdala. But people began confusing the fact that they might not hear anything when they click their amygdala- so I just started calling it 'tickling your amygdala', because its more like a silent mercury switch for most people, like the kind of light switch that doesn't make any noise."

SH: "Well, I don't know if your going to have a woman that is going to be so bold as to point it out to you, but for women, you can very easily imagine, if anything, the lightest touch- almost not even touching touch- can provide the greatest amount of arousal. You know what I mean? If you can understand the other button, it's easy to see the same kind of switch in your head. (laughs) A male brain might not even notice, because the physiology is so different." (laughs)

NS: "My wife has told me exactly the same thing in the way you're describing it. It's very light, it's pleasant, and it's virtually effortless. That's why I like 'tickling' better than the older 'clicking' description."

SH: "I don't know if you'll be brave enough to put that in print..."

NS: "Oh, you'd be surprised."

AMYGDALA TICKLE #5- "Super Synapse Soak"

Take time to relax in a hot tub of water. Soak for twenty minutes and just let go of everything. Add salts or scents to supercharge your forward tickle via olfactory connections to your amygdala. A perfect balance of temperature- Not too hot and not too cold, but Just Right.

Genius Japanese inventor Japanese Dr. Yoshiro Nakamatsu (4000 patents) says, "My best ideas come when I am in water!"

Chapter 8
BRAIN CONSCIOUSNESS PHYSICS

1. ENERGY L.I.F.E. VECTORS

As we've observed, L.I.F.E. is a Line Indicating Forward Energy to, through, and from a living organism- like you.

You – a human being- is a tube, and energy moves towards you, through you, and then the energy leaves you, either as heat, water, carbon dioxide, other forms of radiant energy, or fertilizer.

It looks like this:

Some forms of life are not so much tubes as you and your dog Fido are, but are more like cups or saucers, such as jellyfish, bacteria, and the occasional bit of lichen or moss. So, energy movement might look like this:

You are such a container or tube. In the end the energy that passes through you is what keeps you alive so that you can live to sneeze another day.

Energy movement has direction and magnitude, which is called a "vector".

And in keeping with the balance of nature and the Law of Conservation of Energy, what goes in must equal what comes out, i.e.,
Sustained life and survival in an organism must follow the equation:

Energy In = Energy Out

When this balance is upset, you get sick or die. Poof.

For any living being, energy vectors must move in an overall consistent "forward" direction >>>> i.e., towards Survival.

If the direction of energy into and through the organism is stopped, interrupted, or opposed, the life of the organism will be threatened and diminished.

Like a shark moving through water, energy must continually flow in the right direction, forwards towards survival.

L.I.F.E. is
Line Indicating Forward Energy
Towards survival.

One of the jobs of the Amygdala Compass is to read the movement of L.I.F.E. through you.

An easy to understand metaphor for the balance of energy flow in an organism can be illustrated by a garden hose-

In this illustration, the hose represents an individual living organism. "You" are very much like a hose: Energy in, through, and out.

You eat, you digest, and... you get the picture.

The only real difference between you and your garden hose is that the energy that flows through you also provides you with nutrients that help you grow and replace the parts of you that wear out. A hose is not a living being because it does not grow or reproduce itself and replace its own worn out parts.

Otherwise, I know many people and who have the intellectual equivalent of the hose I water my grass with. They take in energy, it passes through them, and then out from them, but they don't really absorb anything nor grow. They just sit and watch television or send and receive cell phone text messages.

In a properly functioning garden hose, the water spigot is turned on and water (energy) flows from an outer environment source in a forward direction into, through, and finally back out from the hose into the sprinkler and out onto the grass back into the greater environment.

A) If the flow of water stops from turning off the faucet or if it is plugged by a dirt clod or a plug of ice at the input end of the hose- the hose becomes an empty useless and meaningless object, except perhaps as a piece of art you might see in an exhibit at your local art museum.

B) If the flow of water is plugged up at the output end this also will stop the flow of water into the hose, and similarly, it becomes a static container like an unused tube of toothpaste. I hope you do not feel or act like a plugged up tube of toothpaste.

C) If the energy vector and water pressure into the hose is greater than the release of energy leaving the hose, the hose will eventually blow up like a balloon and explode, or the walls of the hose will fail and leak. Then that's the end of the hose. This is somewhat like you when you've eaten way too much for Thanksgiving.

D) Also note that if you are so dumb as to connect one end of the hose to a faucet and then connect the opposite end of the hose to another faucet and turn on both, you would then have energy coming into the hose from opposite directions. This would be very foolish as the direction of water coming from opposite ends of the hose would be in direct opposition to each other and nothing at all would get done, and you certainly couldn't water your lawn. This is very much like the United States Congress.

In any case, if the energy vector into the hose does not equal the energy vector going out, you do not have a working hose- It is for all intents and purposes, a dead end hose.

You probably know some people like that, and they are always asking you for money.

In summary, energy must flow in a forward direction towards fulfillment. Too much or too little energy through the conduit means trouble.

Your amygdala indicates the flow of L.I.F.E., so you know how to go.

Jeff Bailey
Las Vegas Dealer, Ice Cream Master

Jeff has for years helped to run and create the homemade flavors and novelties for a Denver neighborhood home-made frozen custard shop, The Daily Scoop.

Among many other professions in his life, he formerly drove the limousines for major headlining stars as well as having catered to the high rollers of Las Vegas as an expert blackjack dealer. Jeff knows first hand the difference between flushing one's energy down the drain or instead to share it in a fair and even exchange.

NS: "Why do people gamble when they know the odds are stacked against them and always in favor of the house?"

JB: "You know, it's not just the public, but the dealers like myself were always prone to take their weekly check and immediately blow their salary back into the casinos."

NS: "I don't get it- the dealers are in the best position to know from the inside that it's a stacked deck. So everybody, even people on the inside, are all dumb clucks..."

JB: "It's that adrenaline rush, that big rush of endorphins when you win big on occasion. That's all you can think about. Logic doesn't figure into it. You live for the rush."

NS: "How about your present job?"

JB: "I get to interact with the public and I get to have a creative input and output at my job. So I've found a job that not only do I enjoy doing and that I have a talent for, but new people come in all the time and that's really great. I really enjoy what I do now."

NS: "I expect that you never have people come into the ice cream shop who leave crying, unlike Vegas.

"JB: "Yes, and I never have to kick people out because they've had too much ice cream. It's pretty much a win-win situation- people come in happy, and when they leave they're even happier."

NS: "It looks like in your life, you've moved from a job that had as many downs as ups, to a job here where you've got all the ups, but with almost no down side."

JB: "It's all about what you decide is important. I could be making a lot more money- like before- but I wouldn't be as happy as I am now. I've made a lot more money, I've been around and seen a lot of people with money- money doesn't do it. It's what's inside that makes you happy."

AMYGDALA TICKLE #6- "Taste Bud Tickle"

Give yourself a sweet treat. Do it when you're feeling low or just to pat yourself on the back for learning about your amygdala. This is a Fun-Da-Mental tickle, direct from the taste buds to your amygdala.

And while you're at it- take a friend and let the good energy and the tasty tickle flow through you- Any flavor will do.

"A sweet is twice as nice when you share it."

Nils Olaf (Sweden)
Psychiatrist

Nils has been a member of my online Brain Adventure newsletter group for twelve years. He writes:

"Hello again Neil,
Thanks for your kind comments. Quite fun to chat like this across the Atlantic!
In about three months, I will have spent 75 years in this incarnation. I have worked as a psychiatrist in 40 years. For me one of the most rewarding tools has been different variations of "mindfulness", as taught by Thich Nhat Hanh. Some of them are a little complicated, but the shortest one is also the best. It goes like this (translated back from the Swedish edition): 'When you feel irritated (annoyed), smile a little. Calmly breathe in and out three times, keeping the smile.'
I have taught this simple exercise to hundreds of patients, and often forget to do it myself. But every time I remember to do it, it is next to miraculous how effective it is to feel better. And lately I have done it more consistently, and may find myself smiling without remembering why and when I started. I think the power of smiling was discovered thousands of years ago. Now it has even been scientifically proven.
So pleasant forward clicking of the amygdala can be done in different ways!
Greetings, Nils"

I wrote back this response:

"Thanks again Nils,

Why not try this variation:-Smile on your brain!

Smile on your amygdala and frontal lobes! I am certain you will have wonderful results with that variation. [He wrote back later to confirm.]

As always, great to hear from you,

Neil"

AMYGDALA TICKLE #7- "Smile Power"

Close your eyes and visualize a Smile on every part of your body- start at the bottom and work your way up, or just mix things up.

Smile on your toes, smile on your elbows, smile on your spleen, smile on your knees. Smile on your moles, smile on your frontal lobes!

---

2. COOPERATION

Cooperation is not just some arbitrary definition of kissy kissy social behavior, but instead is an exact description of the direction of two or more vectors of L.I.F.E. (Line Indicating Forward Energy.)

Now at first, that might seem like a bunch of baloney New Age crapola granola that I'm dishing out here, but it is actually something quite simple, scientific, and very concrete.

The universe consists of many vectors, some of which enhance your survival and happiness, and others that oppose you. These two opposite movements of energy are expressed as either:

Conflict
or
Cooperation

These two directions of energy can be viewed quite simply:

"Conflict is two or more vectors of energy that subtract from each other and flow in opposition to each other in direction of 90 degrees or more.

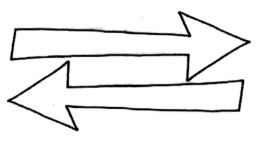

"Cooperation" is two or more vectors of energy that add up and flow in the same general direction of less than 90 degrees.

When you "cooperate" with somebody or something, you go in the same general direction as that energy.

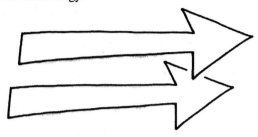

Thus, "Cooperation" is the ADDING of vector energy, and "Conflict" is the SUBTRACTION of vector energy.

You can see everything in your life as energy vectors, either helping you to live, or helping you to fall apart.

Enough food L.I.F.E. vectors going your way, and your tummy remains full and you have enough energy to get through another day at work or play.

If you fail to have enough food L.I.F.E. vectors going your way, you faint from starvation and drive your car across the median on the highway and into oncoming traffic.

It works the same with physical needs as it does with psychological needs.

There are physical acts and vectors which are external.

There are mental constructs and thoughts that are internal.

At the simplest level, you have Cooperation With Self. You need to cooperate with your body and mind so that you act in concert with your personal physical and psychological needs. You need to move in the same direction as your body and mind require to continue healthy operation.

You know what happens when one foot is going southwest and the other foot is going northeast- they don't work so good.

At a more complex level, there is Cooperation With Outside-Self. You need to cooperate with all of those things outside your bodily container, so that you act in concert with your surroundings so that again, you can survive in the context of your environment. No person is an island, even if your name is Joe Oahu. You need to move in the same direction as the environment requires to continue your own operation.

Of note, the reptile brain view is inherently that of conflict, "There are two ways to do anything- my way, and the wrong way."

By contrast, The Frontal Lobe View is infinitely flexible, "Many roads lead to the Pizza Parlor."

For Living Organisms, therefore,

> **"Conflict"** is the movement of multi-vectors away from
> Survival - Happiness.
>
> **"Cooperation"** is the movement of multi-vectors towards
> Survival - Happiness.

The goal of life is survival and the continued flow of energy towards life. Thus, even more simply:

L.I.F.E. is Cooperation.

And guess what? Your amygdala indicates cooperation with pleasure and positive emotion.

Gary Michael
Professional Speaker, Consultant, Fine Artist

Dr. Gary Michael is an artist, a professional speaker, consultant, and author of the book, Get *In Bed With Your Audience,* and *Across A Crowded Room.* He has also published over 200 articles, taught as a college professor (Ph.D. Humanities, master's degrees in both Literature and Philosophy) was a radio announcer, and has acted in television commercials. His house is filled with original paintings and is ringed by a 100 foot fence completely covered with a mural of whales, dolphins, and other sea creatures.

Gary knows the importance of Cooperation- that it is the key to success not only in communication- but in everything you do.

NS: "What key points do you tell people in making their public presentations a success?"

GM: "I don't have a menu or formula. I *listen* to people. I see what their strengths are and build on those, see what their weaknesses are and overcome those- so people make the best of themselves- so people connect with their audience emotionally. The audience doesn't care what you say unless they like you and trust you to start with."

NS: "You just finished writing your first novel; tell me about why you've tried yet something else new..."

GM: "I wanted to see if I could discipline myself to do something that was out of my comfort zone, but that I thought would be fun to try. I wanted to challenge myself. If I don't keep challenging my brain, I fear for its atrophy. Especially with Alzheimer's rampant in my family, I'm thinking, 'Gary, you've got a good brain, you were lucky, so keep using it.'"

NS: "You express yourself in non-conventional ways. Do you think there's too little original thinking in our culture?"

GM: (Laughs) "Brains don't always work in what strikes me as either logical or benevolent fashion. People get locked into their prejudices and their notions of right and wrong. But these same people may have control over their home computers, which utterly confounds me. I can barely turn the darn thing on. I think I actually caught a virus from my computer last winter, and it almost killed me. So, I want to be very modest in my appraisal of other people's intellects.

The city weed police came by here once in response to a neighbor who complained about the natural landscape in front of my house. They didn't even recognize that the native plants that I had growing in my front yard were not noxious and were here long before importing Kentucky Blue grass was installed like a Astroturf carpet everywhere else in the neighborhood.

What there is, is too much thinking that if you don't have what everyone else has- like a manicured lawn- that there is something wrong with you. Life is far more enjoyable when people can express themselves in their own way, however that may be different from your own."

AMYGDALA TICKLE #8- "Listening Light Laser"

Next, when you are with someone- truly Listen to what they are saying- especially when an opinion differs from your own. *Real* listening (starting with *you*) forms the basis for all cooperative behavior and problem solving.

As you listen, see the sound waves enter your ears like a beam of light, traveling through your inner ear onward to your brain- to ultimately tickle your amygdala and light up your frontal lobes.

Henry "Broz" Rowland
Guitarist, Record Producer

Although Broz Rowland has long proven himself to be an exceptional musical talent and music producer, he told me of his days as an insurance adjuster that well illustrated the deadly trap non-cooperative Amygdala Hijacking by the combative reptile brain- and the superior functions of frontal lobes cooperation. We started by talking about the general function of the amygdala.

NS: "So, Broz, the amygdala is like a traffic signal that tells you through your emotions whether or not to proceed with an action: If your amygdala is flashing red with negative emotion, it's signaling for you to STOP what you're doing. If you're getting green with positive emotion, it's telling you GO ahead. Although it's not completely accurate without engaging your frontal lobes, when you add the frontal lobes to the equation, it's a brilliant indicator."

BR: "For me, I see it more like a traffic cop, because it's not only indicating to you that its okay to go, or that you should stop, but because its also directing you to go in a certain direction depending on the need."

NS: "I can dig that. But you know, most places when you get to a traffic signal, and you want to cross the street, there's always one of those buttons on the pole that says, 'To Cross Street Press Button', and you can override the signal and make the light change in your direction when you want to. That's what your frontal lobes can do when you get a red light and you want to go ahead. If you use the smart part of your brain you can take a lemon and make lemonade, you can take a bad situation and judo flip it to your advantage, and tickle your amygdala to switch from red to green."

BR: "Hahah! Most of those buttons are placebos- they're not even connected to the light!"

NS: "Well, I'm talking about one that's actually wired up." (laughs)

BR: (laughs) "But, I do like the fact that its not necessarily instantaneous, but I want to have some conscious input here, so instead of automatically being routed to red light fight or flight, I'd like to have some creative conscious control of this situation and route this puppy into some creative thinking."

NS: "Exactly."

BR: "So this brings up an interesting quandary though- If your amygdala is doing its job in the first place, then why are you messing with it? Maybe it's telling you 'fight or flight baby, get the heck out of here!'"

NS: "Yes, but that's just a do-or-die reaction. Although there are situations where you just react properly like when a truck is headed towards you at fifty miles an hour when you're crossing the street, that's a relatively rare situation.

The problem is when the situation is not life threatening, and your amygdala hijacks your higher reasoning just because you don't like someone's haircut or their accent or because their opinion is different from what you're used to. That happens everywhere you look."

BR: "Oh, you mean because of some cultural thing? We're still taught to act like Rambo Cave Man 'cause it sells more movie tickets, that society hasn't learned to use our big brain instead of our reptile brain. What you're saying is that the amygdala can shunt our consciousness and cause these inappropriate behaviors. You are also implying is that the amygdala can be controlled by a higher intelligence."

NS: "Yep."

BR: "It's like your brain is a house with many rooms. In any of those rooms, there's stuff going on, and your amygdala can stick you in a room where you don't want to be. You get a flat tire, and so you want to get to the flat tire patching tool room. But your boss spent the day chewing you out, and your kid came home with a D- in math. So when your amygdala gets bitten because your tire has suddenly gone flat, it's the last straw. It automatically tosses you into the weapons room, and instead of fixing the problem, you end up blowing your car to smithereens."

NS: "That's what they call it 'Amygdala Hijacking'."

BR: "I've known some people like that.

But let me tell you about something that happened once when I was working as an insurance adjuster. This was after this big storm back east. You want to know how people successfully or unsuccessfully deal with that.

My job was to help people get some money from the insurance company, because believe it or not, I make more money if they get a bigger damage settlement. People typically have the misconception that the adjuster is trying to keep payment to the customer at an absolute minimum, and that is just not the case. I would actually make a lot more money if a larger claim is paid out, so keep that in mind. I had no incentive whatsoever to delay or withhold helping the client get money for damages."

NS: "That's a surprise to me."

BR: "It's actually true, and that's part of the problem.

One instance came up where I could smell trouble from the get go. It was the classic situation of this young couple who were not wealthy, and they just had a baby. The claim was damage to their house, and it wasn't obvious or a given that they were covered by their policy.

There had been some flooding in their house and they were covered for all the damages inside their house. But damage to their roof wasn't covered, simply because there wasn't any new storm damage to their roof. The water had come in because they hadn't maintained their roof, not *because* the storm had done the damage to their roof. None at all. But they still wanted a whole new roof, and insisted that their insurance company pay for it."

NS: "Ah ha."

BR: "I tried to explain that to them, slowly, and very carefully, and in fact that they would be covered for all of the damages inside the house because of the sudden deluge of water that forced its way through their worn out, dilapidated roof. But they were going for broke. They wanted the insurance company to just flat out replace their entire roof, and I couldn't find any evidence to support that. I looked- mainly because if I could have found any and shown it, I actually could have made more money as a percentage of the claim paid out. But I just couldn't find anything to show that the roof had been damaged by the storm."

NS: "Nobody is going to believe that."

BR: "But it's true! Trust me, I was in it to make money, but the insurance company wasn't going to give me a bonus because I'm saving them a payout. It just doesn't work that way, and *it couldn't* because that would be a conflict of interest in my position. I'm not trying to make excuses or make out like insurance companies are saints, but that's just how it is. They can pay for accidents and damages caused by a disaster, but they just can't pay for maintenance a client is responsible for themselves, just because they needed a new roof all along. I tried to explain that to these in a rational way, but they just wouldn't listen."

NS: "Let's make sure I've got this right. They had a worn out roof, and during this big storm, water got into the house because the roof had been neglected. Their policy covered all the damages inside, but it wouldn't pay for a new roof because it wasn't damaged in the storm?"

BR: "You got it exactly."

NS: "You tried to explain that to them."

BR: "I *tried*, but they just went ballistic. The woman immediately went into rant mode when I showed up, and I couldn't get a word in edgewise. She literally started screaming at me like a banshee. But I don't make any decisions at all, I'm just there to be an observer, collect the facts, take photos, and get the evidence."

NS: "You would think she would want to get on your good side."

BR: "Oh yeah, I tried to explain that to them. But they wanted to win the lottery. They got greedy. They were certain I was trying to cheat them, but I'm not making any decisions at all, I'm just seeing what is what and documenting it. The first time I met with them, I couldn't even do anything because she was a total maniac."

NS: "So you're saying that their reason and logic got hijacked by their emotions, they had an Amygdala Hijacking?"

BR: "You got it. So a couple days later after being thrown out of their house, I got them back on the phone and I'm trying to explain to this woman that I'm on her side. I'm trying to find evidence to support her claim that the roof had been damaged in the storm, and that it's actually to my personal advantage to do so, because I could get a commission. I'm telling her, 'I'm here to help you,'"

So, after our phone conversation, she calmed down and I went back to their house a second time. But then her amygdala clicked backwards all over again. It was back to the screaming rant all over again just like the first time."

NS: "It sounds like she figured that if she listened to you nicely on the phone, she'd then get her way in the end. But when that didn't work, she just went back to Komodo Dragon blood dripping teeth and claws all over again."

BR: "Precisely. She couldn't tickle her brain forward, no matter how I explained it. It was like her entire pre-frontal cortex had been lobotomized.

It ended up delaying the whole process of her getting something at all, because I had to go on to the next case. She then had to wait all over again for someone else to show up and take up the claim.

Her payment was greatly delayed and reduced because I was actually trying to help her figure out everything that she was entitled to and get it faster. She got half of what she could have gotten, because she refused to sit down and have a rational conversation with me. My hands were tied. She sabotaged herself because she was just stuck in this 'You are the enemy!' attack mentality."

NS: "Wow. Bit herself on the big toe."

BR: "Now, just the opposite happened with this other case I had in the same storm. This guy had house damage that was not obvious to either him, me or the insurance company. But because he elicited such a cooperative vibe, because he was extremely informative, because he was non-judgmental and he wasn't defensive or at war with me, it worked out perfectly for him in the end."

NS: "So he had this attitude that you were going to work with on his problems together?"

BR: "The funny thing was that his business was that he was actually a consultant, that he comes into businesses as a problem solver to figure out creative solutions. It was his job to think with his big brain instead of his reactive brain.

He had extensive damage in his basement, and he was going to be covered for that, no problem. That's all he expected and he was satisfied with that. But as I'm going over his case back at the office, I turned to one of the guys there with me, and we started looking at what had happened with his furnace and we discovered that he was also covered for this other thing that he never even expected, and at first *I didn't realize either*. But as we looked at it, we found all this other stuff and it turned out to be quite substantial.

In the end, he ended up getting completely taken care of in ways nobody at first imagined. But since he cooperated, and let me do my job and take my time, and because he helped me check things out in a cooperative spirit, we found stuff that otherwise had he been at war with me like the other crazy lady, we never would have discovered.

I was encouraged to pay attention to all the details when I was at his house. I took careful notes, and because I wasn't distracted, I discovered something which he certainly had no idea about and that I would have probably missed entirely if he had been engaged in combat with me.

I think the moral of this story is, don't throw rocks at a stranger. They might just be someone you want on your side. And who knows, one day they might also be someone you want to have produce your next album."

AMYGDALA TICKLE #9- "Cooperation Calculation"
Think about someone you will be dealing with in a casual, personal, or business situation.

1) See how this person can HELP your cause. 2) Figure how you can help this person help you. 3) VISUALIZE this happening.

This will tickle both your logical and left amygdala and your intuitive, imaginative right amygdala.

---

3. AMGYDALA COOPERATION/CONFLICT COMPASS

Your amygdala is a compass that indicates your direction- either towards happiness and survival, or away from it.

And it is inherently a Cooperation/Conflict meter.

If you are to survive and be happy, you must detect which direction your Amygdala Compass is pointing. Most of the time this is an easy matter, but you can be fooled.

Just like a real magnetic compass, there are LOCAL FORCES that can disrupt a compass reading, like being in a strong electromagnetic field, close to electrical wires or in a shielded area. Strong local forces can render a compass reading inaccurate or even change it to read in the completely wrong direction.

A single compass reading might be inaccurate.

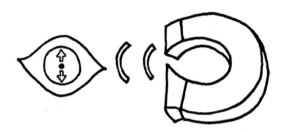

The trick is to see what your amygdala is picking up both locally as well as globally. For that, you need to use your frontal lobes.

Your frontal lobes can compute MANY READINGS from MANY SOURCES, not just what is inside the box you might be presently sitting in.

It can factor in long and short term cause and effect, time, vectors outside your immediate senses, imagination and planning.

In this way, the frontal lobes can calculate an accurate amygdala compass reading and let you know the RIGHT DIRECTION to go, not just for the moment- but for the long term.

Anna "Scout" Wise
Film Maker

Scout has an independent and creative spirit bent in the direction of the visual arts. Her very first own film, *A Stan Needs A Maid*, is a comedic short about the attempted robbery in the house of a pack rat hoarder. Her final film school project, it was chosen as part of the prestigious 2011 Denver Starz Film Festival. This is an admirable accomplishment for a twenty-three year old first time director who had not yet even worn her cap and gown.

Scout has learned well the importance of including- or not including- the input of other points of view.

NS: "You creatively tickle your amygdala with your friends…"

AW: "Film making is a completely natural extension of my personality. I feel compelled to tell stories, because it seems like there is nothing else I want to do. It is who I am. When I do something else, another job that I have to do that's not film, *it's just not fun.* When I finish a film project, when I've finished the work, the color correction, the editing, and it's complete, it's just so gratifying. Whenever I've done something just for the money, it's not the same."

NS: "What was really important in helping you to finish that first film to such a successful conclusion?"

AW: "Oh my gosh. You should always, always, *always* have other people looking at what you're doing as you go along. My film started out as being twenty-eight minutes long, and my friends saw it and told me, 'You have this entire scene in here that you don't need, you've already got another scene that does the same thing.' That made the film so much better, and I couldn't see it for myself. There are other film makers that don't work that way- but I'm never that way. I need the feedback of other people.

When you just look at your own thing by yourself, you cling to stuff because you did it yourself, and maybe you put a lot of work into it. But somebody else, they don't care about what you did as much as they care about whether it works or not. When you're working really hard and long on something, you really need a fresh eye to look at what you're doing."

NS: "Okay, but if you're the director or boss, ultimately, you have to decide what you keep and what you throw away. In the end, how do you know what's right?"

AW: "Oh yeah, well, sometimes people will give you feedback, and you just *know intuitively* that their opinion smells like a bunch of horse poop."

NS: "That makes sense. I guess that's why your amygdala is connected both to your frontal lobes as well is directly to your nose."

AW: (laughs)

Fred Poindexter
Musician, Painter

Fred Poindexter is a highly original artist of many tools and genres. He is guitar player, composer, fine artist, and a wood worker. Fred has written hundreds of songs, and performed thousands of times in front of small and large audiences throughout North America.

Fred's unique perspectives and ability to let go of boundaries and pre-conceived notions is something he shared with his friend Buckminster Fuller, whose geodesic dome home he lived in and was caretaker of from 1975-79.

NS: "What are you listening to inside your head that guides you in the creative process?"

FP: "The artist educates and entertains at the same time. I call it "edutainment". You change the world one painting at a time, one song at a time. Generally, it unfolds from a higher source. It's not me, it's coming through me. That's the way I work.

I don't have a goal in mind. I just show up and start painting, or start writing music. I listen to what I do once I get started. Whatever I'm working on will lead me to another section of the music that works with whatever I started with."

NS: "So you don't have a formula? You live, forgive the pun, *daangerously.*"

FP: "Heh heh. Yeah. Just knowing basic music knowledge, the formula for song writing, verse, chorus, bridge, intro, outro. The biggest mistake that one can make is to over-produce something. You've got to know when to stop, know when it's finished. That's important. Most people know when to stop brushing their teeth, but a lot of artists don't know when to put their brush or pen down."

NS: "You've been to the brain lab with me, decades ago. Any thoughts on this idea of tickling the amygdala?"

FP: "I think it's really helped me. Instead of living in fight or flight mode, you can be more creative by being more aware of the thoughts that go on inside our head. We can go beyond fear and that primitive consciousness and jump into the frontal lobes."

NS: "Obviously you know the conventional rules of making music and art. Do you deliberately break the rules? How do you know you've done something right? What's the feedback?"

FP: "The feedback is internally that I'm happy with it. If it's not quite right yet, I'll feel that it's not quite right. It's done when I'm really happy with it."

AMYGDALA TICKLE #10- "Amygdala Traffic Signal"
Next time you have a decision to make- take multiple compass readings- Green means "Approved!" Red means "Nope!" You can *see* it-
Imagine or visualize a traffic light superimposed over the image of your feedback source- take several readings for accuracy: First, take a Self reading, then take a variety of readings from Others, how they see things.

Ina Hambrick
Hatha Yoga School Instructor and Proprietor

Ina Hambrick has had a career as both a yoga school proprietor and hatha yoga teacher that has continued for over four decades. She may be credited for establishing one of the first, if not the very first non-denominational yoga school for the general public in North America.

NS: "What made yoga so appealing to you?"

IH: "I learned how to go totally within myself, and not care what anybody else was doing or thinking about me. The light bulb moment was when I realized, 'Wow, there are about five or six women in here, and I don't care about what they think of me, or who I am, or what I am.' I had lived my whole life prior to that moment worried about what others had thought of me.

I've always loved the open endedness of my own practice. I don't care for 'one way'. Everybody has their own style. And we allowed for people to do things in their own fashion."

NS: "How did you figure out what was right for *you*?"

IH: "I came to trust, totally, the inner voice inside me, 'Namaste- I respect the inner voice or energy inside me, and I respect that in you.' If I second guessed my inner guide, I'd go wrong. I was so in touch with the energy in myself, that if I had a decision, I'd think about it, I'd meditate, and there was the answer. But if I second guessed that answer, it almost always went wrong.

There was an energy inside me that guided me through everything, and still does. Often there is neither intellect nor emotion. That guidance, it just *is*. It just is *right*. There is certainty there, it is unshakable.

The reason for doing yoga is to have Knowledge of Self, nothing more, nothing less. Every person is at a different level. All our awarenesses and needs are different. Knowledge of Self is very different for each person. And you always keep learning."

NS: "What about meditation? Do you recommend anything to your students these days?"

IH: "I don't teach meditation anymore because I feel that everyone meditates in a different way. I love walking meditation, although some say you can't move while you meditate. I've had some of my highest moments listening to music. I don't want to say to anyone, 'What you are doing is not right.'

I do continue to teach deep breathing (yogic pranayama), and I know that this is especially important among the elderly, and I've seen miraculous results from that."

NS: "I do know the deep breathing is a proven method for relieving stress, and I know it can quickly reverse that fight or flight response."

IH: "You recall that I had been assaulted one afternoon while I was alone at the yoga studio. Although it was a pretty messy sight and I had a fractured skull, because of the conditioning I had been involved with, long hours of yoga and meditation seven days a week, when that attack occurred I didn't suffer any of that immediate or prolonged fight or flight reaction trauma that you would expect. I had no serious repercussions from the incident, it was just an event."

NS: "You know about tickling the amygdala, and something like using the feather visualization to do it. What do you think about it?"

IH: "I'm one of those people who have difficulty visualizing images, so what I do is just *feel it* instead."

4. THE PHYSICS OF MORALITY

As you become aware of your Amygdala Compass emotional feedback, you become aware of Emotional Magnetics and the flow of Cooperative Energy Direction and the flow of Conflict Energy.

The consequence is that sometimes you will move in a direction in conflict with the smaller immediate environment but in cooperation with the Big Picture. This is known as being "A-head of Your Time" or being "an opposition leader".

Example: Although you might live with a family of cannibals, and to oppose dinner made of your neighbors would appear to be in conflict with your immediate family's dinner plans, it would be in cooperation with the bigger global family who did not partake of "prime ribs". Time to leave home.

Conversely, you can make the big mistake of moving in cooperation with the local environment but in conflict with The Big Picture. This is known as succumbing to "peer pressure" or "joining a crazy cult". Cannibals, tyrants and their followers thrive in that kind of situation.

Strange as it sounds, at times millions of people may be moving in the wrong direction in opposition to survival like lemmings high diving off a cliff.

Thus, it is your priority to truly cooperate with the greater forces of L.I.F.E. no matter what other forces and vectors are in your local environment-your own, those immediately around, or at even perhaps a greater level. You brain must see the REALLY BIG PICTURE.

This is what is known as:

MORALITY

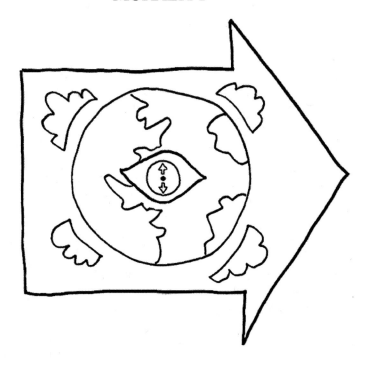

Morality, the overall direction of L.I.F.E. and the ability to cooperate with those vectors moving towards survival can be instantly grasped with your frontal lobes, specifically your *right* pre-frontal cortex brain, the locus of intuition and *global perception.*

Complimentary to the right side of your brain, the *details* of achieving cooperation, which can often be a complex riddle, will be solved with the other half of your frontal lobes, the logic and linear calculations available in the *left* pre-frontal cortex.

You need to use BOTH to truly cooperate with L.I.F.E. and achieve Morality and remain happy.

Balance. Your amygdala remains tickled by your frontal lobes, no matter what anybody else is doing or however many votes they got in the last election.

AMYGDALA TICKLE #12- "Brain Balance Question 1 & 2"
Ask two questions:
1) "What do I want for myself?"
2) "What does the greater universe want for me?"

Logic it out and *Feel* it out. The true, balanced, and harmonious frontal lobes solution will be indicated by a positive emotion answer to BOTH questions simultaneously. Take note and remember. Then smile- ☺ You can have your cake and eat it too.

Chapter 9
BRAIN RADAR and NON-LOCAL CONSCIOUSNESS

Brain Radar has it roots before written history, a concept that can be traced to the very idea of who or what was behind all that thunderous noise way up in the sky before it rained on your Neanderthal mullet haircut.

"Ugg. I think there's something bigger than me."

The mistake of modern man has been to relegate the idea of non-local consciousness as being simply a matter of un-provable superstition and religion. However *truly* modern man has come to realize the potential reality of non-locality in quantum physics.

Quantum physics has shown us that electrons appear to jump in zero-time from one electron ring to another, without going through anything- they just appear in one spot, and then magically appear in a completely different location instantly, not much different from Captain Kirk instantly beaming down from the Starship Enterprise to the pleasure planet Risa for a little vacation time.

Einstein stated that nothing can move faster than the speed of light. But in 2011 the OPERA Project (Oscillation Project with Emulsion Tracking Apparatus) in conjunction with the CERN particle accelerator on the French-Swiss border- seems to have demonstrated that neutrinos do just that; move faster than light. (Of course, this 'proof' ping pongs with competing projects.)

This is all relevant because the entire idea that our brain can know something beyond the reach and speed of our senses and/or know things that exceed the speed of light may be entirely real.

"About 80 years ago, scientists discovered that it is possible to be in two locations at the same time—at least for an atom or a subatomic particle, such as an electron. For such tiny objects, the world is governed by a madhouse set of physical laws known as quantum mechanics. At that size range, every bit of matter and energy exists in a state of blurry flux, allowing it to occupy not just two locations but an infinite number of them simultaneously." (Discover Magazine)

Although most of us can't seem to be in two places at the same time- as surely many of us would like to do on some occasions- there are in fact impressive and corroborated stories of saints even in contemporary times, such as Neem Karoli Baba, who could accomplish that astonishing feat.

However, your more common un-saintly brain can do what your two feet alone can't manage with the ability to perceive and detect particles or waves that carry information from anywhere "out there" to where ever we are "in here".

This is clearly possible even with the most ordinary little bit of imagination. Imagine now, sitting there reading this book, you are standing on the moon looking back at Earth. Voila. And faster than light to boot.

Beyond that rudimentary demonstration, our brains often appear to utilize subtle energy beyond the detection of even our most sensitive instruments, even though many hard core skeptics wearing blinders will deny this till their dreary dogmatic dying day. Everyone has experienced such a thing at least once in their lives, and many experience it regularly.

Sometimes it may only be in knowing when you know someone is going to call you on the phone, and with no apparent warning or reason, seconds later, the phone rings- and it's them.

We all have stories that go far beyond the remote probabilities of pure chance, of finding ourselves in exactly the right place at the right time, knowing something that was impossible to know, far beyond a random guess or chance.

This is Brain Radar at work.

Michelle McCoskey
Computer Software Technology (Apple, IBM)

Michelle resides in Marin County, California and has been part of the high tech revolution for decades, having been on the ground floor as part of the designing team for the first software used in Apple and IBM personal computers.

NS: "How did you first learn about amygdala tickling?"

MM: "I heard you the first time you were on the Art Bell Show. I read your first books back then."

NS: "Back then we did that first mass brain focus thing, when they had that serious drought in Florida."

MM: "It does work. It seems like all of these different experiments... we have our own group here that we call 'The Intentioneers' and we go out into what we call 'The Field', and put out intentions for everything. There were three or four of us that were busy tickling our amygdala after we heard you, and we still do. I also do Tai Chi brain integration exercises.

My other really good friend who read the books and is in the group is into brain physiology and psychology. Her professor was at Tulane and did the brain manipulation stuff [Dr. Robert Health, Tulane University, direct amygdala stimulation]."

NS: "What's your formal educational background and your occupation?"

MM: "I have a bachelor and master's degree in Social Sciences from Indiana University. I was a music archivist for the music school there. Then I got into computer software and worked with John Draper, who was known as Captain Crunch. We put out the world's first word processing software for the Apple II computer and for IBM."

NS: "Oh, tell me about that..."

MM: "My partner and I bought an Apple I..."

NS: "An Apple ONE?"

MM: "Yes. That was the early days. We bought it at a hardware store."

NS: "Wow. People are familiar with the Apple II, which was the one that really took off and was the one that looked like a small typewriter, but you're talking about the first Apple computer... You probably had Steve Jobs and Steve Wozniak's fingerprints on the thing..."

MM: "Yeah, and I've known all those guys. This is like 1979 and the 70's in San Francisco was quite bizarre."

NS: "Hey man, groovy baby. Did you hang out with the Grateful Dead, man?"

MM: (laughs) "A little bit. Jerry Garcia's Mountain Girl, who was with Ken Kesey, was in my circle of friends. [Mountain Girl was Garcia's second wife, and was one of Kesey's 'Merry Pranksters, *The Electric Kool-Aid Acid Test by Tom Wolfe*]"

NS: "How did you end up with an Apple I?"

MM: "It's karma. Or my astrological chart, or whatever. One of my distant ancestors crossed the Delaware with George Washington, and he was a general and a governor..."

NS: "I actually have a five foot wide hand painted copy of that picture of Washington Crossing the Delaware on my wall. My grandfather commissioned it done directly from the original hanging in the Metropolitan Museum of Art."

MM: "My relative is the one standing up. [next to Washington] His name was General Arthur St. Clair. Both sides of my family were teachers and farmers, and pretty liberal kind of folk. So in the 1960s I did all of that kind of mind expanding stuff, you know. And... okay, I can tell you this...

When I was about four and a half years old, I lived at my grandparent's house in this little town north of Indianapolis. One day I was on the outside porch and I saw these three people come down out of the sky- two men and a woman."

NS: "They were floating down?"

MM: "Kind of coming down on an escalator that you couldn't see. They were coming down into this field across the street. And I thought they were Jesus, Mary, and Joseph with his coat of many colors. But I thought they were a bit scary and I didn't want to go with them."

NS: "Well, I wouldn't have gone either."

MM: "They came into the house looking for me, and I hid behind the refrigerator. My grandmother was in the back yard hanging up clothes. I saw that the woman made a mark on a little chair. You know how kids have a little chair and table? And then they left, and went back up and they were gone. So I knew right back then that reality was bigger than what everybody else was saying it was."

NS: "Did you go look at the mark?"

MM: "Yes, and the mark was there for years, and I used to show it to my little friends."

NS: "What was the mark? I mean, does the word 'crayons' mean anything to you?"

MM: "No. It wasn't my Crayolas. It was just a dark oval. It wasn't a symbol or anything. It was a black mark. A few years ago I talked to some of my friends from way back then, and they remember the story and they remember seeing the mark. The whole thing happened, whatever it was.

Then another time I saw someone in my room and I thought it was Abe Lincoln's *first* wife."

NS: (laughs)

MM: "Now, I don't know how a four year old knows that Abe Lincoln had a first wife, but he did have a first wife. And she told me 'Don't put your shoes on the edge of the bed,' and I've never figured that out."

NS: (laughs more)

MM: "Surrealism has always been involved."

NS: "So you haven't been diagnosed with some sort of mental problem?"

MM: "No, no, I'm extremely competent. (laughs) I've always thought that they were hallucinations when I was little, but now I know that they were just some sort of information that I still can't quite figure out."

NS: "Well, even if they were 'just' hallucinations, they were pretty darn good hallucinations."

MM: "Exactly. I did psychedelics way back in the '60s, and these visions were nothing like that. Those were pretty colors and lights, but these visions were real, they were reality."

NS: "While you're on the subject of psychedelics, you know, everyone is worshiping Steve Jobs this week [after his death, Oct. 2011], and I have to say that it's well known that he took a lot of psychedelics."

MM: "They all did. My favorite was Steve Wozniak. John Draper was a 'phone freak' person and he was friends with them."

NS: "Oh, sure- their first project was building one of those boxes that would allow you to hack in and make free long distance phone calls. These are the same guys who now make computers that cost twice as much as anyone else's and who scream if anyone tries to steal their own stuff..."

MM: "I actually have one somewhere."

NS: "They made like a hundred of those, and then they realized, 'Hey, we can make a lot of money building little devices with buttons, and that led to the first Apple computer."

MM: "Yeah. So I got involved in this group called 'The Apple Core', and twelve people showed up- Wozniak, and Jobs, and Draper were there, and my partner and I hooked up with Draper who ended up writing the code for 'Easy Writer', the first software for the Apple.

Jobs couldn't stand Draper because he was so 'uncivilized' and wild. But Draper got along well with Wozniak, who seemed to be the most heart-full person in the group."

NS: "What about Bill Gates? Did you know him?"

MM: "Yes. I rubbed elbows with him, not knowing he would end up being one of the richest persons in the world. But those computer guys were all so geeky, because all they would ever talk about was about the engineering and 'Oh, and you can do this with this and...' I'm a much more metaphysical person; my technology is 'Brain'."

NS: "You're using the circuit board between your two ears."

MM: "Exactly. That's why I like clicking my amygdala."

NS: "Well, the brain can run circles around anything. The brain made the computers. So computers are only a little tiny slice of what the brain is capable of computing."

MM: "Exactly. There's so many different realms the brain is capable of working in- most people don't even have any comprehension. We're still at a stage where we've hardly got a handle on its energy."

NS: "The brain is an infinity machine."

MM: "The other visualization I do besides clicking my amygdala is 'Polishing Your Telomeres' You know the ends of your DNA strands are called 'telomeres', and they kind of wear out as you get older- that's just a gross description of it."

NS: "Like a fraying at the ends?"

MM: "That's kind of it. So you just polish them to keep them from wearing out. So I just do that visualization too."

NS: "And the DNA is contained in all of the mitochondria in all of our cells- so you're visualizing going down to the cellular level, and the sub-cellular level and addressing the DNA."

MM: "Yeah. I address the DNA, 'Hello DNA'."

NS: "So back to your group- So you would get together with a few people, and you would click or tickle your amygdala together- then what?"

MM: "Well, we would visualize that and then would share whatever we experienced or saw. We would experience altered realities, and we would do problem solving together. We would go in and work on activist things, and healing things. Basically it was all for healing- society and people."

NS: "Do you still do this?"

MM: "Yes, off and on. I still do the simple amygdala tickling, although I don't hear a click or anything- I just feel a little spring, a spiral twist- a spring that maybe goes "Boing!""

NS: "Oh, that's exactly one of the things that I wrote about in *The Book of Wands*, that sound that some people hear when they get a good Brain Pop."

MM: "Yeah, it's more like that. I feel like its energy above my eyebrows going out."

NS: "I've drawn a picture of that in another book- that the energy isn't just circulating around inside your head, but it's actually going out from your forehead and your frontal lobes in a forward direction."

MM: "Yes, that's basically how it goes, and how I visualize it."

NS: "Can you describe an emotion that you associate with it?"

MM: "It's like a happiness, but more like a lightening, a positive feeling. I think of it as clicking, but it's more like a tickling thing."

NS: "So, how often do you tickle your amygdala?"

MM: "I do it whenever I think of it, which sometimes could be not for days, and then for days on end. Where I live is across the Golden Gate Bridge, and west. So I do it driving home. I do it a lot when I drive, looking up at the open sky."

NS: "Do you have to be sitting home, or when there's no noise?"

MM: "Oh, no, no, no. I can be in the grocery store. I actually think I do it more when I'm doing things, like my brain integration thing."

NS: "Describe that."

MM: "It's tapping on the outside of your head in a certain pattern. It's a Tai Chi exercise that you use your right middle finger, and it clears things. But I saw a relationship to this Tai Chi exercise, and a metaphysical relationship on the head called 'The Square of Saturn', and then tickling your amygdala as well. Then there's another one called 'Beating of the Heavenly Drum' where you use your fingers on the back of your head."

NS: "So, it sounds like you see a convergence of all of these various disciplines. There's a similar brain exercise that I learned at the brain lab. I talked to an osteopath in Belgium who tickles his amygdala who does something similar with the hands."

MM: "When I tickle my amygdala I would just have to say that I've made things a little easier and cleared the energy in my head."

NS: "Okay. That's a short term result. What about long term effects?"

MM: "One would have to say that it enhances equanimity. It has a stabilizing general mood effect."

NS: "Like balancing your car tires for, um, a smoother ride, better gas mileage, and longer wear?"

MM: "Yeah. It definitely puts things in alignment. It seems to make your brain waves synchronize better or something like that."

NS: "Describe your life before and after you started doing all of these different brain things?"

MM: "Now I'm better off. All of my life I've had a progression of my spirit and soul, and this is one of the pieces that has made things better. All of my life I've known about these kinds of things, but this has been a useful tool in keeping stability and mental coherence, this is one of those things that has come along to help overcome or calm the 'monkey-mind'."

NS: "There are all kinds of things that you can do to better your life- there's a plethora of self-improvement techniques out there. What does tickling your amygdala offer you that you don't find elsewhere?"

MM: "Well, number one, it's simple. It's not dogmatic. It doesn't have a ritual involved with it- it doesn't have a 'have-to' involved with it. It's a simple thing to do. And to me it has results of coherency and harmony."

God helps those who help themselves, and there it is- you're helping yourself."

AMYGDALA TICKLE #13- "Glial Group"

Two heads are better than one. And three are better than two.
Get together with a friend, or two, or three, or more.
Tickle your amygdala together- however you choose. See what happens.
Try it once. Twice. Or more. Then see what happens.

BRAIN RADAR – SCIENTIFIC INVESTIGATION

The idea that you can know or even manipulate something that is happening far beyond any conceivable way of knowing or touching it has been called by a variety of names: ESP, paranormal perception, pre-cognition, clairvoyance, clairaudience, telekinesis.

This is the stuff of Magik and legend, the stuff of Wizards and Sorcerers. In some circles, it's seen as being Looney Bird stuffing.

In many other circles, it's a way of life and quite real.

To understand that our brain operates at levels beyond our conscious understanding is simply to understand how our brain works at the most fundamental level in the first place.

To begin with, much of what goes on in our body and brain occurs at an unconscious level.

Take for example, the regulation of your own body temperature. How are you doing that with your conscious mind and brain? The manner in which your brain regulates your metabolism and body temperature is a rather remarkable achievement, something that even a baby accomplishes remarkably well. How about your hearing? How to you manage *that* trick?

When you jump at hearing a loud noise or laugh at a funny joke, these are processes that occur without our conscious willing of it to be so.

If we would had to have a rational explanation for every thing we accomplish as Homo sapiens, we would have never outlived our Neanderthal cousins 43,000 years ago. We would have vanished long before the Dodo bird.

Further, if we were conscious of everything we are doing at every moment, we could never get anything done. Our minds would be flooded with a crippling universe of sight, sound, and intention. We could not focus on any single task, even flossing our teeth.

This is simply why the nature of our brain is to filter out so much of our experience and relegate these tasks automatically to sub-conscious and unconscious juggling deep within our brain.

History has demonstrated that we often are clueless about the way things work. One year we are convinced the world is flat, and the next year we are sure we understand the world shaped like an overstuffed ping pong ball. But despite how much we think we know or don't know, your You-niverse continues to spin on its merry way as sure as the Earth circles the sun.

So, many things that we witness, see and hear, we will perceive and know exist without knowing why. Just because we still don't understand exactly how gravity keeps the Earth attached to our solar mother's apron strings doesn't mean that we can't trust that the sun will rise each morning as we have come to expect.

Often we find ourselves consciously face to face with something new for which we have no simple rational understanding. You find that you have miraculously arrived at the right place at the right time without understanding why or how.

If you are like most people, you've had hunches that have proven correct in the end, with no reason to explain how you knew such a thing. You probably have known what another person was thinking or known what they were going to do before they did it. And you were 100% right on target.

You may have denied or dismissed such inexplicable knowing as just a coincidence. But this is a natural gift of your brain. This is your miraculous Brain Radar at work.

When you consciously recognize this as an actual and real process, you can begin to rely on it and allow it to occur when it is needed most- *if* you know how it operates.

The way in which the neural structure of our brain can utilize mental "wormholes" in consciousness that allow an individual to escape the confines and limitations of local of three-dimension time-space has been examined by Roger Penrose (Rouse Ball Professor of Mathematics, University of Oxford) and Stuart Hameroff (Departments of Anesthesiology and Psychology, University of Arizona).

They propose a practical and neurological explanation for the ability to see beyond our simple five senses that they call Orch OR. They describe this in their paper entitled *Orchestrated Objective Reduction of Quantum Coherence in Brain Microtubules: The "Orch OR" Model for Consciousness.* (See: http://www.quantumconsciousness.org/penrose hameroff/orchOR.html)

Theoretical physicist Amit Goswami Ph.D, (Professor, University of Oregon Institute of Theoretical Science, 30 years, ret.) often cites several experiments that demonstrate this phenomenon of Brain Radar.

In Goswami's words, "...Psychic phenomena, such as distant viewing and out-of-body experiences, are examples of the non-local operation of consciousness... Quantum mechanics under-girds such a theory by providing crucial support for the case of non-locality of consciousness." (A. Goswami, *The Self-Aware Universe: How Consciousness Creates the Material World,* 1993.)

Goswami cites an experiment by neurophysiologist Grinberg-Zylberbaum, and described in *"The Einstein-Podolsky-Rosen Paradox in the Brain; The Transferred Potential, Physics Essays 7, (4), 1994."*

In this experiment two subjects meditated together for twenty minutes. A total of seven pairs of subjects of both sexes, with ages from 20-44 years participated in the study. After meditation and while maintaining their "direct communication" (without speech), they were placed in electro-magnetically shielded chambers separated by 45 feet. Both subjects were connected to EEG instruments and 100 random flashes of light were presented to Subject A. Subject B was not told when the light was flashed for subject A. The results indicated when subject A was stimulated, subject B's brain responded in exactly the same way, even though the second person was in no way being stimulated with the signals or being given cheats on the sly by the doctor in attendance.

The same experiment was replicated by Peter Fenwick and collaborators in London in 1998, and again with the positive results in Seattle at Bastyr University in 2004. There, neuroscientist and researcher Leanna Standish, ND, Ph.D. gives her own name for Brain Radar telepathy, and gives it a more scientific moniker calling it "Distant Neural Signaling".

Online, Torbjorn Sassersson has compiled a comprehensive list of more than 120 scientific papers, abstracts, laboratory experiments performed at universities, and including peer review papers and scientific journal publications that prove the existence of Brain Radar in its many forms. This list can be found in his web site page titled *Evidence of Paranormal phenomena, ESP, PSI, Remote Viewing, Neuroquantology, Quantum Nonlocality, Near Death, and Out of Body Experiences.* (See: Evidence of Paranormal Phenomena, http://torbjornsassersson.com/research/vetenskapliga-bevis-parapsykologi.shtml)

Even the entertainment field gives us at least one genuine exponent of remarkable Brain Radar in the personality of The Amazing Kreskin (his legal name). At the age of 76 he still dazzles audiences with a talent that he describes as a natural god-given gift, and not a trick of any type. I can vouch personally, as I was myself present at one of his shows to which I took my mother for her birthday one year.

To my utter surprise, he chose me for one of his demonstrations. I can vouch that I spoke to no one before the show, nor do I walk with a limp.

In the middle of the performance, he announced my father's birthday. "Does October 17th mean anything to anyone here?" he said to the large audience. I was the only one in the auditorium who stood up. He then thought for a moment, and not only did he tell me I had surgery on my right knee from a sports injury as a kid (true), but he gave the first name of my orthopedic surgeon, whom I hadn't spoken to nor had even thought about for thirty years. Given the chances of identifying a previous injury on a body capable of any number of injuries, I'd say that was a darn good trick.

Kreskin calls himself a "mentalist", and doesn't describe his ability as ESP. He realized his unusual talent as a child, when his brother hid a penny in their house one afternoon, and that he was able to find it nearly instantly, without rhyme or reason, nor without sneaking a look where the coin had been placed. For decades, Kreskin has offered a prize of $50,000 to anyone who can prove that he uses confederates, advanced set-ups, tricks, or electronic or other listening devices to read the minds of persons in attendance of his wildly entertaining yet apparently real demonstrations of Brain Radar.

In half a century, despite ample opportunity across the globe, no one has ever succeeded in proving Kreskin a be a fake. His prize has remained uncollected.

Perhaps an even more compelling case has been documented over the course of fourteen years by Nancy Talbott, who heads a team of a dozen professional university and business scientific investigators at B.L.T. Research in Cambridge Massachusetts. Among other notable cases she has studied are the astonishing abilities of thirty-two year old Robbert van den Broeke.

Among recent tests given to Robbert was a double-blind test in which he could duplicate a hand-drawn image such as a boat or other scene or a random set of numbers drawn around on a sheet of paper. Further, his drawn facsimile was not vague or approximate- it was an exact duplicate in which he reproduced every element in exactly the right size and shape to the source image.

Robbert not only could do this from a separate floor of the building from which the picked drawings would be selected at random, but he could do this within a three minute time limit, and additionally with the restriction that the investigators themselves would not know the content of the selected drawing, i.e. a double-blind test. His score in doing this? 100% success.

Notably, Robbert repeatedly yet cheerfully admonishes anyone looking into his feats of Brain Radar to say that they are all well within the capabilities of every person.

Nancy Talbott
President, BLT Research Team, Inc., www.bltresearch.com

Nancy Talbott is president of the BLT Research Team, Inc. a non-profit professional U.S. research group consisting of multiple credentialed scientific consultants and a varying number of field-workers. The BLT Research Team's purpose is the discovery, scientific documentation and evaluation of physical changes induced in plants, soils and other materials sampled or recovered from crop circle sites world-wide.

NS: "My book deals with using the rational and intuitive frontal lobes as opposed to the reactive reptile brain. Can you illustrate how you approach your work in a scientific and intelligent, and open minded manner?"

NT: "Well, anyone pursuing a scientific understanding of something new must utilize standardized scientific methodology, insofar as it is possible. As soon as John Burke, Levengood and I had teamed up I developed a routine sampling protocol for the various field workers who were collecting the plant & soil samples from crop circles in various countries. Levengood developed a standardized methodology in his laboratory and he also carried out some control studies to double-check some of what he was finding in the crop circle samples. The 3 peer-reviewed papers BLT published during the 1990s would never have been accepted for publication by peer-reviewed scientific journals if we had not followed accepted protocols."

NS: "So, the fact that you've followed these protocols says not only something about the work that you're doing but it says something about what it means to use your brain in an intelligent manner."

NT: "I'm lucky! (laughs) Those of us who are able to apply both approaches- rational and intuitive- to the understanding of new or as-yet-unknown phenomena are fortunate. Many of the people who have produced insightful work in the so-called 'anomalous' arena (UFOs, Bigfoot, crop circles, whatever) have had to be able to use both sides of their brains. They've had to think both within the current paradigm as well as "outside the box." I think it's almost impossible to make a truly solid contribution if you're not able, at least some of the time, to venture outside the realm of conventional thought.

Another very real problem facing researchers examining the currently 'unknown' is how do you reassure people not involved in such work, and quiet down the fears that many of the questions you ask raise in people? Because I think a very basic fear of the unknown- just because it is unknown- prevents many people from even thinking about asking questions."

NS: "I was thinking exactly that same thing two seconds before you said it. I mean, the way the fear gets in the way of effective thinking."

NT: "Well, when all of the various media outlets, like TV, movies, the internet, and sensationalistic print media imply or straight-out tell you there might be danger in pursuing certain ideas, many people accept that this information as authoritative. And if such people are frightened enough, their brains can shut down altogether. Many people have very limiting concepts and ideas which are very often supported primarily by fear, especially when the habit of always accepting the pronouncements of people who are 'authorities' is learned when they were kids."

NS: "What strikes me about crop circles is that they are intrinsically beautiful designs..."

NT: "Yeah, a lot of them are. But a lot of these fantastic formations are made by people- please remember that. It really is likely that many of the elaborate designs are mechanically-flattened by people, lots of them by young men in response to the ideas they hear the tourists talking about in the English pubs at the end of the day. I'm afraid that some of the most popular designs in the UK every summer are man-made. But, that being said, there are lovely formations every year, in the UK and elsewhere, that are the real McCoy."

NS: "In the film that I saw this morning [Crop Circles-Quest For Truth], a great many designs- some very elaborate- one that comes to mind was a huge spiral which appeared near Stonehenge in just a twenty-five minute period of time during the afternoon... ["Julia Set", July 1996]"

NT: "Based on the examination BLT carried out on that one, we'd say that one wasn't man-made. In particular, we found extreme germination abnormalities in the seeds taken from the crop circle plants, abnormalities which existed not only in the flattened crop inside the circles, but also in a 'spillover' effect up to 300 ft. outside the flattened circles. Levengood felt that the energy system which created that formation was not only very strong, but also very complex. This was also his opinion about the huge July 7, 2007 'necklace' formation in East Field."

NS: "That was a very beautiful design..."

NT: "In recent years many British crop circles every summer are very intricate. Many crop circle enthusiasts as well as some of the crop circle 'researchers' are absolutely convinced they are all genuine, but come to these conclusions based on personal ideas which have not been substantiated by any formal study. One of the recent 'themes' attributed to some of the recent British formations is the idea that some circles have represented a Mayan 'message', a substantiation of apocalyptic events on the horizon. Other formations have been interpreted as representing a wide array of doomsday scenarios.

Regarding the Mayan theme I am very close to 100% certain that the first formation which was interpreted in this way, a formation near Silbury Hill in 2004, was man-made. An elderly American man whom I know well was in the UK that summer and went into that formation on the first evening it was found. He was hoping he might get some photo anomalies, but instead got another kind of shock. He had no idea that the formation was only half-made, but after only 15-20 minutes inside the formation he watched several young British guys march up the tramlines and into the circle, carrying planks and ropes- and these guys immediately set about finishing the design in a very methodical manner.

Two other 'croppies' were also visiting the formation at this same time as my American friend, two young Englishmen the American knew slightly, who had also come to try and get some photo anomalies. All three of these men then watched the hoaxers as they added to the original design, working quickly and as if they knew exactly what they were doing.

My American friend was a bit intimidated because one of the hoaxers walked up to him and told him he was not supposed to tell anybody about what he was seeing....so he left after ½-hour or so. Apparently the two young British visitors also left soon after. And in the morning, lo and behold, there was this lovely 'Mayan' design- now completed.

I tried to give this information to a number of crop circle enthusiasts I know well, but nobody in the community- and this is some of the researchers as well as tourists- wanted to hear about my long conversation with the American man who had called to tell me all this when he got home. Personally, I figure that if the 'second part' of the formation was without question man-made- observed by three separate people as it was being completed--then the 'first part' almost certainly was also man-made. But I met with no success in trying to share the eyewitness information I had. And in a few cases I almost felt the people I was trying to inform wanted to stone me!" (laughs)

NS: "What do you look for that distinguishes a real crop circle from a man-made one?"

NT: "There are multiple plant and soil scientifically-tested changes which can distinguish the difference, but only two of these changes are visible to the naked eye in the field- and one of them is not always present. The most reliable plant change, which has been tested in hundreds of formations in many different countries over a 12-year period, is apical node elongation. I mean here the top node of the plant, the first node beneath the seed-head. When this node is visibly elongated you can be very close to certain that the event is genuine. However, this elongation- or stretching- is sometimes so subtle that the only way to determine it as a certainty is to actually measure hundreds and hundreds of the circle plants and a similar number of control plants from outside the circle, a considerable distance away, and then run a statistical evaluation.

In some cases the elongation of these top nodes is one-hundred or two-hundred percent above the controls- and in these cases you can easily see the difference between the samples and the controls quite clearly. In England, this apical node elongation is often not as dramatic as we have seen in other countries, thus requiring a lot of work to determine a solid statistical difference. Generally, we require at least a statistically-solid 35% elongation or greater to consider the change seriously; and when the apical node elongation is this slight we must also be able to demonstrate some of the other abnormalities by carrying out germination studies in the lab, soil examination, etc. to conclude 'authenticity' of a given formation."

NS: "What's the second test for a non-made made circle?"

NT: "The second scientifically tested plant indicator of authenticity- although it is not always present in real circles- are holes at the nodes, which we call expulsion cavities, and these are usually seen in the nodes farther down the plant stems."

NS: "And that's from the steam escaping from the stalk of the plant..."

NT: Yes, this is a part of Levengood's plasma vortex hypothesis. He believes that one of the energies involved in creating real circles is a heating agency, most likely very intense and very brief (nanosecond) bursts of microwave radiation. When the microwaves hit the plants they immediately turn the plant stem's internal moisture to steam... and since steam expands, it stretches the young, more elastic nodes at the top of the plants as it escapes. Farther down the plant stem the external fibers are much tougher and they do not stretch- so, as the steam builds up, it simply blows a hole to escape at these lower nodes."

NS: "I know you have been working on the crop circle mystery from another angle also, one that is much harder for science to deal with. I'm talking about your work with the Dutch medium Robbert van den Broeke..."

NT: "Yes. What I've learned by working with Robbert for going-on fifteen years now is that a consciousness of some kind- and I don't know if it's my consciousness or his, or Carl Jung's 'collective unconscious' or somebody or something else's- is involved. After all these years, and the hundreds of totally off-the-wall events I have observed personally while with Robbert in Holland, I no longer have any doubt that a consciousness of some kind is directly involved with the crop circle phenomenon.

If you take the time to carefully read through my reports on Robbert's case on my site you'll begin to understand why I say this. I have seen two crop circles form right in front of me while in Holland with him. And he has seen many, many form over the years. He is also the only person in the world, so far as anybody knows, who 'knows' when new circles in his area are getting ready to appear. He can feel them coming- and for seventeen years now he has always been correct. While at home he will draw a picture of what the new circle is going to look like and he also sees in his mind's eye where it will be. And when we drive to the place he has 'seen', there is always just what he has said will be there. Sometimes the new circle is still forming as he arrives, and he and whoever has driven him to the field often also see a wide range of light phenomena in or near the new crop circle field.

In addition, other anomalous events are occurring around him all the time, events we label differently such as poltergeist activity, and remote viewing, and out-of-body experiences, 'mind-reading' and telepathy, healings and contact with the deceased, ET and UFO encounters, and even the materialization of physical objects. Not to mention the thousands of utterly bizarre and inexplicable, clear, photos taken with many different people's cameras and in front of dozens of witnesses of so many strange things we're running out of categories by which to call them.

There is a consciousness of some kind involved in all of this, and I think these anomalous events which people currently label separately are all aspects of this one incredibly complex and absolutely not-yet-understood consciousness. I'm not totally certain whether it's Robbert's consciousness somehow being projected, or mine, or some manifestation of a global human consciousness- whether it's angels, ETs, inter-dimesionals or what. I just don't know for certain.

But Robbert is certain it is a huge cosmic consciousness totally external to him and the rest of us, and further, that it is a loving, and positive, spiritually-based energy. This is what he has always said, all these years, and what he absolutely believes. And I have no reason so far to think he's wrong- and many reasons to strongly suspect he is right. I am certain that the consciousness behind all these bizarre events is the same consciousness involved in the creation of the real crop circles."

NS: "My observation in the research for this book points to two distinct forms of consciousness: One I call the 'Little Me Eye' which is the common, egotistical, self-centered- and I don't mean that in a negative way- but the way in which we look at the world through our own eyes as individuals. But then there is this other level of consciousness that I call the 'Big Magic I' spelled with the capital letter 'I', which is a quantum consciousness, a collective non-local consciousness- brains connected together into a mind that is beyond any single individual experience."

NT: "That's what Carl Jung called 'the collective unconscious' I think. That's definitely a possibility. The energy behind the circles could be all of those bigger things we've mentioned, or it could be a sort of inter-dimensional reality. I don't know exactly what 'inter-dimensional' really means, so I can't comment too intelligently on this. But the fact that there is a consciousness, something which is completely aware- in Robbert's case and in the circle phenomenon- has made itself clearly apparent to him and me so many times that I can no longer say to myself, 'Oh my gosh, that was just a weird coincidence!' No. I'm a big girl now and I've seen too much. A reasonable person can fight the new knowledge for only so long. I had no choice but to bite the bullet and quit trying to convince myself that what's going on is 'coincidence'."

NS "You're talking about experiencing a whole other level of reality that we can experience as individuals, but that goes far beyond what we are as individual persons with this little package of typical awareness inside of our brain."

NT: "We're all connected, we're all connected. And not just to each other, but to this amazing 'energy'. As soon as we can get that through our thick heads, the sooner we can learn to behave in a much more intelligent fashion. What I've learned over the years is that there are other intelligences and consciousnesses 'out there' which very likely may have been there all along, and that many of them may have been interacting with humanity from the get-go. I also suspect that there may be a very wide range of agendas among these other life-forms, some of which from my vantage point are the 'good' guys, others that aren't. And, further, that whatever situation that you find yourself in- whether it's with humans or some other unknown thing- if you aren't comfortable, if you don't feel the kind of respect for your needs which you normally require from your everyday involvements- then get the heck out of there!"

NS: "What you are saying is directly related to the subject of this book, which is the emotional feedback compass of the amygdala..."

NT: "Ahhh, yes. There are a lot of people all over the world who are trying in different ways to learn about the larger reality. The crop circles are just one approach, one way of trying to open oneself to a larger, more accurate understanding... just one enticement from 'The Big Kahuna in the Sky' to encourage us to try harder, to be more grown up, to recognize the challenge to grow and, so, fulfill our greater potential.

These other consciousness are a bit seductive. They're saying 'Come on, let's play! Open your eyes... this is fun!' And one way they seduce us is by creating lovely, mysterious designs in our fields, interesting and curious patterns which aren't scary at all, but which just sit there inviting whoever wants to come in to do so and sit a spell. Many other anomalous events are scary- Bigfoot would scare me and we know that UFOs scare a lot of people. But crop circles are just sit there enticingly in the field, not doing anything, just offering a pretty place for people to hang out in on a beautiful day. It's a totally different kind of introduction to the possibility that we still have a lot to learn. To me the circles seem to be the most elegant idea that whatever it is that's out there has come up with to gently encourage people to take the next step."

AMYGDALA TICKLE #14- "Left Brain Science Search"

Your left frontal lobes looks at details from a rational and logical perspective.

Find a controversial subject, perhaps one that is often the target of skeptical scorn.

Then READ and READ SOME MORE. Look for the real scientific investigation beyond the conventional and dismissive ignorance.

COSMIC CONSCIOUSNESS DNA

In order for any kind of communication to occur between individuals, there must be something shared in common between the two. It can be visual signals, auditory signals, language, touch, or anything at all which is a common thread tying both separate entities together.

If there is nothing in common, communication can not take place. This would be akin to trying to order Indian food speaking Swahili at McDonalds.

If there is something in common, then information can be swapped through the sympathetic vibrations of like elements in each individual.

This sympathetic vibration can be easily observed on a piano. For example, if you strike any key, such as a "C" and cause one C string to vibrate, all of the other C strings will begin to vibrate to some degree as well without you physically touching any of them. This is known as sympathetic vibration.

If Brain Radar is a reality, if the ability to communicate with distant non-local objects and individuals over time and space is a reality, then there must be some type of sympathetic vibration happening between you and that rock on the dark side of the moon, or between you and that furry two-headed pink alien pet shop owner circling Sirius, the "Dog Star".

What could that possibly be?

Charles Darwin has provided us with a clue.

Darwin proposed and did a pretty good job giving evidence that all living creatures on Earth have a common ancestor. This was basically outlined in his book of 1859, *On the Origin of Species*.

His theory was later borne out in the discovery of DNA by Watson and Crick in 1953, the molecular helix arrangement of just four simple amino acids that is present in every single living creature, from the smallest microscopic bacteria to the largest blue whale. DNA can be found in every lonely and dumb bit of tundra moss and in The Smartest Person in the Whole Galaxy (probably perceived by yourself as you).

DNA is the thread, it is the one thing in common shared by every living creature, one cell or one-hundred trillion cells. And it can be traced back as far in time as we can go. It is seen in the fossilized hairy footpads of a giant Jurassic mosquito, and earlier.

So the question for us and for the possibility of Brain Radar remains: What could you or I possibly have in common with a talking mushroom in a far distant galaxy, or for that matter, how could we possibly know what someone whom we don't even know is doing on the other side of the planet- as Brain Radar purports us to allow?

How could we possibly transcend the limitation of time and space with our consciousness and our brain? *We would have to have something in common with everything in the known universe and have the ability to sympathetically vibrate with it.*

This "thing", the medium by which Brain Radar connects all things, I would propose and name, is Cosmic DNA. It is the tin cup to tin cup string than allows the sympathetic vibration between you and that spinning top set in motion by a little kid playing unknown in his little thatch roofed cottage deep in the Siberian forest.

If you go far, far, far back in time, you will come to the origin of the universe- the moment just before the Big Bang, when all that existed was a singularity- a single condensed point of everything. At the moment of the Big Bang this singularity expanded and split into parts from which all of the elements and substance of the universe was made.

You are literally made of the same stuff of stars, and planets, and you are made up of the same elementary elements and particles from which everything else was formed, because all is joined in unison with a common cosmic ancestor.

If you go backward to the point of singularity, there was a single sub-atomic particle or thingy from which everything was recombined and added together to make up every star, planet, and every creature that sits on every slimy rock.

After the Big Bang came helium, and from that, all of the other elements were formed. But of course, there was something even simpler still that preceded helium, and these were the sub-atomic elements from which helium congealed through nucleosynthesis which occurred even closer to the moment right after the Big Bang Bunny left his bunny tunnel.

That same elementary sub-sub-atomic thingy remains in each atom in every atom in the universe, in the same way that biological DNA is found in every living creature. It doesn't go away. It's the fundamental building block of the universe.

That sub-sub-atomic thingy is Cosmic DNA.

It is this most elementary of the elementals, the Cosmic DNA which expanded via the Big Bang and recombined like a google of Cosmic Bricks to make everything that exists in the universe. It is everywhere.

It is the thing by which Brain Radar communicates via sympathetic vibration throughout the universe. It is the thing in common between all things that allows your brain to vibrate and KNOW what your long lost celestial cousin is doing below the surface of Mars, without shaking his paw.

As previously noted in this book, the ability and neural structure for the human brain to detect signals on the scale of sub-atomic particles and neutrinos is outlined by Oxford mathematician Roger Penrose and Stuart Hameroff of the University of Arizona in their paper, *Orchestrated Objective Reduction of Quantum Coherence in Brain Microtubules: The "Orch OR" Model for Consciousness.*

Brains talk to each other, regardless of space, in the medium of the commonly shared element, Cosmic DNA.

In 2011, atomic physicists have found evidence for an elementary sub-atomic particle which may illustrate the existence of such a Cosmic DNA. This particle is named The Higgs Boson (pronounced boh-zon) particle, nicknamed by Nobel Prize-winning physicist Leon M. Lederman to the consternation of some scientists, "The God Particle".

As Lederman later explained, this nickname has been subject to some misinterpretation, "The publisher wouldn't let us call it the G-ddamn Particle, though that might be a more appropriate title, given its villainous nature and the expense it is causing."

The Higgs boson is a subatomic particle proposed by the British physicist Peter Higgs in the 1960's. It is thought to uniformly endow everything in the universe with mass, thus its nickname referring to its omnipresence throughout the universe. Further, the confirmed observations conducted on more than one occasion indicate that this neutrino sub-atomic particle moves faster than light.

It may very well be that one day in a future discovery we may happen upon solid proof of Cosmic DNA itself (besides your own experience of course), which may make the speed of the Higgs Boson appear as if it moves like a sedated snail by comparison.

Steven Snyder
Piano Technician (for Steinway Pianos, Carnegie Hall, Lincoln Center, Stevie Wonder's principal technician)

The story of this interview is something quite remarkable in itself. Steve had ordered several of my books from me, and I ended up talking to him on the phone for a short while. As soon as we began to talk, I realized what a remarkable and unique story he had to tell, and that it was not only a story about tuning a musical instrument, but something much more important about the nature of consciousness.

His story is perhaps not so much about tuning a musical instrument, but more about tuning the human brain.

NS: "Tell me what you do?"

SS: "I tune and repair pianos, I rebuild them, and I'm considered an expert piano action regulator."

NS: "How did you learn how to do this?"

SS: "My father did it, and I started as a kid. Then I studied with top notch regulators throughout my career. I worked for Steinway in New York City in the '70s after I graduated from Boston University."

NS: "You've got to be pretty good to work for Steinway."

SS: "Um, yeah. Pretty much you do. I'm considered an expert in the field, you might say."

NS: "Plus, you've been doing this your whole life. But you also mentioned to me that you worked for Stevie Wonder. Tell me a bit about that."

SS: "When I was at Steinway, Stevie had his recording studio on 8th Ave. in New York City, and I tuned his pianos at that recording studio. He was getting ready to move all of his operations to L.A.. He had tried piano tuners in New York, Chicago, and Los Angeles- and he liked my tunings the best. He offered to set me up and help me get established in Los Angeles if I would move. So I did, and I worked for Stevie for the next ten years, from about 1978 to 1988."

NS: "I have to ask you a question- You said that he liked your tunings, and a lot of people probably ask themselves, 'Well, couldn't you just get a guitar tuner and a screwdriver?' I mean, what is there to tuning a piano? Of course, I know... but a lot of people might not."

SS: (laughs) "Well, from my point of view, which is admittedly pretty odd, I view each piano as a living being. It's more than just the sum of its parts- it has consciousness. The way I view everything in the universe is that everything has consciousness- even your car, a chair, or a table. It all has consciousness.

When a piano is assembled it has its own unique sound and tone, and that comes from that consciousness present in that instrument and everything, all the consciousness that went into it to manufacture it. You take two pianos that are manufactured the same day with the same materials by the same people in the same factory and each one will sound different. They'll be completely different because of the consciousness present in that instrument.

So what I do besides tuning- which I do by ear with a tuning fork- I get feedback from the consciousness in the piano, and I work with that feedback to find the very best tuning for that instrument. So, it can't really be duplicated- if you get a really good tune on a concert Steinway grand you can't just go to another concert Steinway with the exact same thing and get the same results. It just won't sound right because each piano is unique. Each consciousness in each instrument is unique, and to get it to shine and come out, it needs to be done differently for every instrument.

I thought I was pretty hot stuff- I worked through college for four years, and I worked with my father. When I first went into Steinway they were looking for tuners at the time, and I thought I was pretty hot and that I would get hired. But they turned me down flat, because my tunings were not good enough for them.

So, I worked for Macy's for a while, and then Sohmer, and Baldwin. I kept going back to Steinway and kept taking the test, but they told me, 'Well, we know you can hear, but your tunings aren't up to our standards.'

After a while I got to know everyone over there and they finally said, 'Okay, we'll give you thirty days and will give you our best in-house tuner to train with, and he'll teach you Steinway tuning technique.'

You see, I had learned a temperament from my father, who was a pretty good tuner, but he wasn't really good enough for professional work. He was really good for the home, and he could do schools, and churches, but when it came to recording studios he wasn't up to those standards. I carried that tradition forward which reinforced all kinds of bad habits. I had to drop all of those bad habits and learn everything that Steinway taught me from the ground up of how to tune all over again.

But at the end of about twenty-eight days of training I was so confused, and paranoid, and upset that I could hardly tune at all. The test day was only a couple of days away and I was all upset and worried. I went to sleep one night and I had this dream-vision kind of thing where I was lying in this filed of wild flowers. It was springtime right after a rain and all of the wildflowers were pungent and the smells were fantastic.

I was lying on my back in the dream among all the flowers looking up at a crystal clear blue sky. Instead of clouds in the sky, all of a sudden all of these puzzle pieces started drifting down. They were puzzle pieces of bits of strings, and piano pieces, and tuning levers, and tuning mutes and felts and everything that I had been working with and struggling with to make sense of to do a quality tuning. They were just floating down out of the sky. Then all of these separate pieces all came together into a solid picture.

Once I saw that picture, I knew how to tune a piano. I knew what Steinway was trying to teach me, and it all came together in a large concept of this is how it's done.

The next day I went in, and I could do it. I passed the test at Steinway and I started to work for them."

NS: "It sounds like you're describing something that you can't even put words to..."

SS: "Yeah. It just came together kind of like- I didn't really choose to become a piano tuner; it chose me to do this kind of work.

After that is when I really hit a high level of performance when I started doing Carnegie Hall and Lincoln Center, and all the performance venues and the recording studios in mid-town Manhattan, and I really started to go."

NS: "I'm absolutely certain that most people that go to a concert or listen to a professional recording of a piano, they haven't given even two seconds of thought to the piano tuner. Heck, *I haven't*, and I'm a professional musician! I haven't even given it that much thought until this moment that there is a difference between somebody like you and the guy who tuned my piano for thirty years.

Now what is even more interesting, is that you told me the first time that we talked, that you were completely deaf for the first part of your life..."

SS: "The first five years."

NS: "Was it total deafness?"

SS: "Yeah, total. I had overgrown tonsils and adenoids. And what that did was to block my hearing canals, and cause them to fill with fluid. It sealed out sound, and the tympanic membranes in my ears couldn't move and I couldn't hear a thing.

If I had lived in the middle-ages, I'd still be deaf. But the problem was discovered and it just took a simple operation to correct it. ·

When someone has one of their senses that is lessened, then often nature compensates by heightening the sensitivity of another sense. In my case, for the lack of hearing nature compensated by developing my eyesight, where I could see what people might call 'the invisible realm'.

I could see the material world, but I could also perceive hyperspace and parallel realities, and all of the beings living in those other realities.

This isn't that unusual because a lot of children see these realities. But what happens is that these kinds of kids are made fun of, and the culture drums it out of them so they learn to deny that aspect of themselves. So, they shut it off instead of letting it develop naturally. But it my case, since I wasn't talking and because I wasn't hearing anything, the most active environment for me to put my focus on was hyperspace and parallel realities, and the beings who lived there.

Those beings started communicating with me and started giving me my education right then and there on the nature of reality, and what kind of world we live in, and what human beings are doing, and what they are and what different body types live in those dimensions. So, for the first five years of my life, those were the kinds of beings I interacted with."

NS: "Well there goes all your credibility. And mine with it. Thank you very much."

SS: (laughs)

NS: "Actually, that's something a lot of people have experienced in my end of the neighborhood. Tell me a bit about your own experience tickling your amygdala, because you've indicated to me that you've had some real results with that as well."

SS: "I read about it about a year ago and I've been working with it intensely ever since."

NS: "What techniques have you used to do it?"

SS: "I tried all the different ways. I do it every day and I've incorporated it into my routine. The focus of my whole life is spiritual- so everything I do, including my work tuning pianos, is supportive of this kind of lifestyle I have.

I've incorporated the amygdala stuff with my daily routine. One thing that I do for example is to use a tuning fork that's tuned to 523 Hz. I'll put that tuning fork on my forehead and seek to have that aid in tickling the amygdala forward. I'm also putting it on the top of my head as well, I use both places."

NS: "Okay. The top of the head would correspond to what is called 'the crown chakra' in eastern philosophy, and at the forehead is where your pre-frontal cortex is- where key processes for higher consciousness and advanced brain functions are integrated.

Do you tickle your amygdala when you're out doing your daily activities as well?"

SS: "No, actually not. I do this more like a formal meditation. But I will also do this throughout the day on occasion during my free time. When I'm out in the world I focus on what I'm doing, and I don't want to have an experience where suddenly I'm incoherent and can't function because of the experience I'm having." (laughs)

NS: "What do you experience when you use your tuning fork to tickle your amygdala? I'm guessing that you've noticed some result or you wouldn't keep doing it."

SS: "It gets my brain and my whole being more focused in that area, like a key switch to increase brain usage. I have noticed that there's a real connection between gratitude and love, and coming from the heart and the amygdala flipping to the frontal lobes.

I think that this whole connection between the heart and the pre-frontal cortex is something that is profound. For me, in my experience, they work together. If the heart is functioning with the pre-frontal cortex in a greater capacity then there is more of a chance that I have coherent thinking instead of reactionary survival thinking."

AMYGDALA TICKLE #15- "In Tune Tickle"
Get a tuning fork and make a tone with it, then hold the handle against your sphenoid bone- on your temple, outside the amygdala. Try the same thing holding it against your forehead- outside your frontal lobes.

BY THE WAY...

True story:
As a small minor aside, it was my very own Brain Radar which allowed me to predict the very existence of the Higgs Boson God Particle at the age of seventeen, only a senior in high school with a B- grade average. Perhaps this was not unlike Einstein himself, who discovered relativity while still working as a lowly patent clerk.

It happened this way...

My best friend Robert Kennedy and I, (yes, that was really his name, but this R.K. was a violin playing teenager in Denver, Colorado), had been experimenting as was our occasional practice at that time, with the neurological effects of ingesting 3,4,5-trimethoxy-ß-phenethylamine, commonly distributed among our school buddies as "chocolate mescaline".

We had been inspired by the sociological research of the indigenous native Indians of the Sonora desert of northern Mexico, as written by anthropologist Carlos Castaneda, as reported in his books, *The Teachings of Don Juan: A Yaqui Way of Knowledge*, and *A Separate Reality*.

So it was on that fateful day that Robert and I had altered our brain chemistry and had reclined in a supine position on the grass at one of our favorite city parks, Cook Park in southeastern Denver.

With our ever expanding awareness, we gazed into the distant rich blue sky above us to meditate on the ever changing cumulus cloud formations.

At one point, completely unexpected and to our amazement, we saw one cloud metamorphose into a clearly recognizable figure that we had been familiar with since childhood.

One cloud stood out alone from the others, and it was unmistakable:
The cloud looked exactly like Bozo the Clown.

At once, the realization dawned upon us, communicated as it were as we leapt far into the distant future via Brain Radar and Cosmic Intelligence.

We realized at that moment, and said to our teenage selves, 'There we have it and can see it- Bozo *is* God.' And we henceforth declared ourselves as Bozonians and claimed our new faith.

So it was, through our own brains upon the wings of Brain Radar that we had traveled across the vast distances of time and space. We drew the same conclusion forty-two years before the CERN physicists detected and which

Leon Lederman also knew, that the Higgs Boson was in fact "The God Particle".

Okay, granted, the spelling is a little different.

Nobody is perfect.

AMYGDALA TICKLE #16- "Cloudbusting"

A fun brain game is to go outside on a sunny day when big puffy cumulus clouds are floating by.

Pick one cloud among many others and first tickle your amygdala to get in the right mood mode. Then focus your Brain Energy on the chosen cloud, as if you could vaporize it with your intention to do so. Watch what happens. It typically takes 3-5 minutes.

Coincidence? Brain Magic? Disappearing Cloud Precognition?

More info at www.neilslade.com/cloud.html

A FULL UNIVERSE FULL OF PENCILS

Another elegant solution about how non-local communication and consciousness takes place has been suggested to me by Jim Casart, a friend with whom I've been playing music with every Monday night for the past fifteen years.

Jim is a CPA as well as a guitar player. Leave it to a musical tax calculator to come up with a more efficient explanation than me. I'm not sure if this is his original idea or not, but regardless, he explained it to me one night in between "She Loves You" and "I Want To Hold Your Hand"...

"All you have to do is see the universe as being full. Astrophysicists already know that most of the universe is made up of dark matter, so it's not that far a stretch to think that the universe is completely full of something- that there is not really any such thing as empty space.

Just because we haven't detected it or have a name for it yet, doesn't mean that everything isn't connected to everything else. Just see the universe as *full*.

So if that's the case, then whenever something in the universe moves, it moves everything else everywhere- like a pencil.

If you move the eraser end of a pencil, then the point end moves at exactly the same time and in exactly the same direction.

If everything in the universe is connected, then if I move my hands apart here in my apartment, then everything else that is connected to my hands also moves.

Maybe that's how non-local consciousness works, and in that way, there's not even any delay at all.

149

So Neil, now that I've explained how the universe works, let's play another song..."

Tom Meyers
Osteopath (Brussels. Belgium)

Dr. Meyers has developed an approach towards healing which improves prefrontal blood flow, and thus facilitates to get a person out of the fear based reactive brain into the cognitive brain through direct physical stimulation of the exterior cranium.

In this way, Dr. Meyers suggests through his own observations and experiments that the inner workings of your brain can be directly manipulated by touch on the outside of your head, which is connected to the most inner part of the brain through connecting tissues- skin, muscles, bone, then the brain itself- similar to the way in which Jim Casart has suggested that everything in the universe touches everything else.

NS: "Can you explain what you do in a professional capacity?"

TM: "I'm an osteopath. That is a manual therapist who works holistically or globally to address a patient's ailments. We use not only muscular/skeletal techniques, but also soft tissue manipulation. We work with the body to regain a fluidity, so it can better heal itself. We are really catalysts of the healing mechanism that is already present, to allow a person's body to better focus on the essentials."

NS: "One thing that caught my attention when you first contacted me through email quite some time ago, was that you've integrated this very simple idea of amygdala tickling in with your specialty. Could you explain that?"

TM: "As soon as I started to work as an osteopath, my patients who initially came to me for their neck and back problems- which they received relief for often very quickly- they began to comment to me that most remarkably that they could also concentrate better, they could make better decisions, what was foggy became clear..."

NS: "This was after a standard treatment?"

TM: "This was after one of *my* standard treatments. And they would remark to me that they never had this extra mental bonus after treatment by any other osteopath.

So, I was trying to help them as much as I could for their other physical ailments, but the bonus was that they received something more. And this was more on a psychological-behavioral level. They told me, 'We have less stress, and we can cope better with stressful situations after a treatment with you.'

As I heard this over and over again, and with the comment that we don't experience this anywhere else, then I really had to examine my work to understand what I was doing differently from anyone else to make that difference."

NS: "What did you find that you were doing?"

TM: "I found five points of movement associated with the cranium that were abnormal in certain patients which I had adjusted without anticipating any psychological result. Only when I had gone back into my records to see how these patients had been treated differently from the others did I make the connection between my treatment and what people were telling me, without any prompting on my part.

When I went online to try and discover what the connection might be between the psychological result I was seeing and the neurophysiology, I then was able to make a connection between the cranial structure- such as the sphenoid bone [the temples] and the respective internal structure of the brain adjacent to that region. My feeling was that my manipulation of the sphenoid bone was having an effect on the amygdala, which resides on the interior of that cranial plate.

It's interesting, when you look on a search engine and look at photographs of stressed persons, what part of their head are they touching?"

NS: "They're touching their forehead..."

TM: "...Or their sphenoid bone."

NS: "They're putting their hands on the side of their head, or they're holding their head in the front..."

TM: "It's a natural reflex that you've never been taught to do when you feel stressed. You're touching that zone when you feel stressed.

Adrenalin narrows the blood vessels, and causes an increase in areas activated in fight or flight..."

NS: "Which causes a decrease in blood flow to the pre-frontal cortex."

TM: "Exactly. There is a decrease in blood to the pre-frontal cortex and to the neo-cortex in general. In fight or flight, more blood flows to the reptilian core of the brain, and less to the periphery.

In fight or flight you get less glucose and less oxygen and less nutrients to those zones in the advanced parts of your brain, and less potential for clear thinking.

Robert Sapolsky at Stanford University has shown us what happens when brain blood flow is chronically cut off as a result of chronic stress. If there is an impediment to blood flow in the brain, in particular to the hippocampus, if the hippocampus doesn't get enough blood flow then it shrinks, it impairs memory, and you also create a continual production of cortisol which results in depression and burn-out.

There's a whole negative chain reaction because of an interruption of blood flow to certain parts of your brain."

NS: "In simple practical terms, in modern life we are very prone to elevated levels of stress, either work related, or relationships, or economically. And the result of that stress will be impaired memory and impaired problem solving, and will result in depressed mood."

TM: "That's correct.

So, I had the points, and a year and a half later I realized that those points were reflex points that you use your hands to stimulate yourself when you're in a stress moment- you're healing yourself with your hands. But it's not conscious- you're doing this with your subconscious. But it's the most natural thing in the world.

Now I'm teaching people to do this consciously, because if you are aware of it, you can apply it and help yourself in healing more readily."

NS: "When I'm talking about tickling the amygdala, I'm usually speaking of an internal thought process. But you are suggesting that you can use your hands to affect something quite similar."

TM: "That's correct. I've been teaching this to people for more than a year, and I've found that everybody can do it."

NS: "How are people responding?"

TM: "I'm seeing that people are absolutely amazed at what they see they can do for themselves and what they can do on others as well. In other words, you can do this for yourself, but you can also do this for others."

NS: "You're using physical movements of the hands that correlates to the visualization technique of tickling the amygdala using imagery and imagination. Is that correct?"

TM: "Yes. Also, when you touch and place your hands on the forehead, I believe this stimulates the pre-frontal cortex in a similar way.

This is what I am seeing and hearing directly from my patients. It has been remarkable to me."

AMYGDALA TICKLE #17- "Gentle Tickle Touch"
Simply place your hand or your finger tips on your forehead to bring positive energy into your frontal lobes and thus tickle your amygdala.

Do the same thing on your temples, the area immediately in front of the top of your ears, on top of the cranial sphenoid bone.

BRAIN RADAR GOES PHYSICAL

Brain Radar can not only be seen as mental perception and psychological phenomenon, but it can also be seen to extend into real physical manifestation in the body and even into physical reality that extends beyond one's personal body.

In 1969 at the invitation of Dr. Elmer Green of the Menninger Foundation, Swami Rama, an East Indian yoga master, was a consultant in a research project investigating the voluntary control of involuntary states. Later joined by a Dutch man by the name Jack Schwartz and others, these people participated in experiments that helped to revolutionize scientific thinking about the relationship between body and mind. These persons clearly indicated the existence of precise conscious control of autonomic physical responses and mental functioning, previously thought to be impossible.

Under these scientific conditions, Swami Rama demonstrated the conscious and willful ability to stop his heart from pumping blood for up to 16.2 seconds and to produce an 11-degree difference in temperature between different parts of the palm of his hand. In another demonstration using only mental power, he caused a fourteen-inch aluminum knitting needle mounted on a shaft five feet away to spin.

Rama was observed by a team of scientists and wore a tightly fitted foam and Plexiglas mask over his face to prevent any air current from his mouth or nose from affecting the movement of the PK needles in the telekinetic experiment. He was also completely covered and wrapped further with a cloth over his entire body to prevent any other bodily movement from affecting the needles.

Rama claimed that he could cause the energy from his heart chakra, one of seven yogic philosophic centers of consciousness located in the human body, to glow as to be visible to the naked eye. After a skeptical challenge from a physician to prove this, Rama had the doctor point a Polaroid camera at him while he did exactly as he claimed he could. The result was an un-retouched photo, created on the spot in front of witnesses long before the age of Photoshop. The photo shows a fourteen inch wide area of light emanating from the heart chakra area. Josh (*Beyond Biofeedback*, Elmer and Alyce Green; "Walking With a Himalayan Master".)

Jack Schwarz, a survivor of the Nazi concentration camps, repeatedly demonstrated the astonishing potential of the human brain in his lifelong efforts to encourage brain self-control through spiritual exercise and mental concentration. He was founding director of The Aletheia Association and The Consciousness Research Foundation, Grant's Pass, Oregon. (http://www.holisticu.org/)

Born in Dordrecht, Holland in 1924, Jack Schwarz distinguished himself as an internationally recognized authority on voluntary controls and human energy systems. He was a subject, researcher, and consultant at major biomedical and life science research centers in the United States and Europe. His efforts often paralleled examples given by yogis in the Far East. He found that his unusually heightened powers of mind and brain began initially after a spiritual vision he had while unconscious after a tortuous beating by his prison camp guards.

In 1975 Schwarz demonstrated his astonishing mind-over-body abilities on the nationally televised *Mike Douglass Show*. Wearing regular western clothes, he rolled his shirt sleeves up and thrust a twelve-inch darning needle through his biceps, without bleeding, without experiencing any pain, and smiling the entire time at the astonished studio audience.

Schwarz did this repeatedly at request for lab researchers at the Kansas based Menninger Foundation. He was able to stop and start copious bleeding at will, control his heart rate (stop his pulse), hold lit cigarettes to his arm with no pain or permanent skin damage. The cigarette burns ranged from simple red marks to blisters on different occasions. Within seventy-two hours all trace of burns disappeared. Although he had been doing these kinds of demonstrations for years, the researchers remarked that "The skin on Jack's arm is as smooth as a baby's."

All of his puncture "wounds" closed immediately, and were completely healed and completely invisible between twenty-four and forty-eight hours. For added emphasis, he also had a ritual of rubbing the needle around on the dirty floor with his shoe before demonstrating.

Jack's range of vision exceeded what is normally considered to be visible light. As a consequence, he saw the band of electromagnetic fields which surround every organic body. Based upon seventy years of observing the energy fields, Jack Schwarz could describe the physical, emotional, mental, and spiritual conditions of people, and the degree to which they use their unique energy potentials. (http://www.neilslade.com/Papers/Rama.html)

After his research with Rama, Schwarz and others, Dr. Elmer Green went on to be founding director of The International Society for the Study of Subtle Energies and Energy Medicine, aka ISSSEEM, now permanently located in Arvada, Colorado (http://www.issseem.org/).

For the past 23 years, ISSSEEM has published conclusive studies, scientific papers, and double-blind laboratory experiments demonstrating the detectable existence of subtle energy systems, i.e., Brain Radar in all its many forms.

But there are some very recent cases of remarkable individuals who have brought Brain Radar into full physical expression, and done so under repeated and rigorous scientific study, as were those individuals tested at the Menninger Foundation.

Most westerners are familiar with the famous film quote, "I see dead people..." uttered by the adolescent boy in the 1999 film, *The Sixth Sense*. with Bruce Willis. Among his other skills of Brain Radar perception, the previously mentioned Robert van den Broeke not only can "talk" to dead people around him, but give him a camera- preferably one belonging to another person- and he will then produce a picture of such a spirit instantly viewable on the camera's own LCD screen with no chance of any computer software or camera manipulation possible.

But perhaps as mind bending or more so were the abilities of John Chang as observed over many years by Dr. Lawrence Blair, a world respected anthropologist who has spent decades exploring the various island countries within Indonesia.

First looking for a medical cure for Dr. Blair's brother Lorne's eye infection while on one of their expeditions, they came upon Change who had the nickname of "Dynamo Jack". Chang was able to produce electricity with his own body that he used as part of his medical treatments. "DJ" described this energy as "Chi" energy which he could focus on a patient's ailments, but which he could also employ for demonstration purposes: To set newspaper on fire with his bare hands, to thrust any ordinary bamboo chopstick through a thick solid table top, catching rifle bullets with his hands, as well as to actually light LED bulbs held in his finger tips. "But such feats are just the beginning," according to Chang, and are just the outer manifestations of the really interesting work, which he says is internal to the brain.

Reflecting the inherent interconnection of these forces to the frontal lobes C.I.C.I.L. and amygdala, Chang stated that "We must be very aware of our emotions, otherwise these energies can be dangerous. They cannot be used to show off in public or to cause harm to another."

A thoroughly modest individual, Chang's demonstrations were additionally and carefully observed after a thorough medical examination. This included the use of a metal detector, having stripped of all of his clothes, as well as having been done at a random remote location under the supervision of highly skeptical and independent third party witnesses including a physicist from Albert Einstein University, a medical expert, and a scientific investigator. (See http://www.youtube.com/watch?v=5dpBgq_OOHM)

Most importantly, what is the point of all of these "magical" demonstrations and what do these remarkable people all have in common? Chang sums up the common thread and flatly states "We all have undreamed of powers waiting within us, natural and life affirming abilities simply waiting to be released by focusing the innate energy of the mind and brain."

YOUR OWN BRAIN RADAR

Although these well documented examples of Brain Radar illustrate this power in those individuals who can call upon dramatic demonstrations of it at will, such abilities are obviously not necessary for achieving even the most modest amount of personal fulfillment and happiness. They are merely examples that Brain Radar is a real force, and one that can dramatically alter the course of one's life.

What is key is that this immeasurable power of the brain is something that exists within the capability of every person, and that it is a completely natural function of every human brain, and that it will occur *spontaneously* **and effortlessly as needed.**

Rather than seen as a consciously controlled supernatural power, Brain Radar, on the other hand, is a naturally occurring guidepost for one's actions and behavior. It is a natural Normal-Paranormal function of a balanced brain in homeostasis that occurs spontaneously without any additional or special effort. It is S-ESP, Spontaneous Extra-Sensory Perception that happens spontaneously and serendipitously as real and meaningful Magic.

It occurs simply when there is a balance of the focused Doing abilities of the left frontal lobes, and a deliberate access of the Being nature of the right frontal lobes.

When this balance of the brain occurs, it allows for the spontaneous creation of Brain Radar- indicated by positive emotion, pleasure reward indicating movement in the right direction- picked up by your brain's built-in amygdala compass needle.

The most important thing in accessing Brain Radar is knowing what kind of thinking process allows it to occur within your brain:

Brain Radar is a force that manifests through Passive Activity.

Passive Activity is "effortless effort", to do something with a light touch.

Imagine shooting an arrow into a target. You put your arrow in your bow, you aim, *then you let go.*

The arrow flies, and meets its destination.

You don't carry the arrow to the target and push it in.

Brain Radar is like taking a deep breath, and then you allow an exhale, naturally, and without forcing anything. You let nature take its course.

It's not the product of doing nothing, but the product of doing nothing special. It occurs as the result of a perfect balance between Doing, and Being.

Brain Radar cannot occur when your consciousness is preoccupied with the limited vision of your reactive reptile brain. You cannot perceive Brain Radar with The Little Me Eye

This tiny part of your brain cannot tap into Brain Radar. The sensory reactive part of your reptile brain only compute what is one inch in front of your nose, what concerns only You You You, the part of your brain that can only calculate Me Me Me Consciousness.

The self-centered part of your brain is like being shut up in a small box without windows. Everything inside that box is just about your own little world.

The fundamental purpose of the reptile brain is simply to react, to grab for yourself, or to kick back and resist what is not you. The reptile brain computes 100% *competitive* consciousness, fight or flight, attack and counter-attack behaviors. It controls your basic survival reactions, and is obsessed with fighting the "It's Me or You!" battle.

How could Brain Radar possibly function in such an environment, when Brain Radar is all about communication BEYOND self?

Brain Radar is a product of sufficient integration of the frontal lobes. Brain Radar is your consciousness connected to those parts of your brain that sees all around you, into the limitless infinity that exists from your feet and outward past the end of your street to beyond edges of the Milky Way Galaxy.

Brain Radar operates at instantaneous faster than light speed, it is the computation of quantum non-local consciousness data.

Your Brain Radar is a Quantum Train of Thought, a vehicle that laughs at the limitations of ordinary mundane ideas.

Brain Radar works when you tap into the Big Picture, when you see things with your Big Magic I, the 360 degree vista in four, five, and eleven dimensions, far beyond what you can comprehend in words or even vision, past what you have not even dreamt yet.

When you stop thinking about yourself separate from everything, and when you realize you are connected to everything, and that everything is a part of you and you are a part of everything, then Brain Radar waves begin to flow through the ocean of consciousness. Brain Radar travels upon those quicker than light neutrino quantum highways that allow you to see all sides and all possibilities at the same time.

Brain Radar is bigger than you. It is your brain computer connected to the You-Niverse Wide Web, to all the other brain computers connected to the Universal Energy System wherever they are, at instant quantum jump connection speeds.

And there is no monthly charge, or sign up.

To connect and use your Brain Radar connection is simple:

You just tickle your amygdala, and you are on board.

Elizabeth Slowley
Massage Therapist

Delightful and unexpected seemingly "magical" physical manifestations of amygdala tickling are not at all uncommon among persons who regularly engage their frontal lobes in a deliberate and focused mode of helpfulness.

Skeptics would dismiss such events as "coincidence", but veteran amygdala ticklers know better- that the more they engage their frontal lobes, the more frequent such so-called "coincidence" occurs.

One such common story was emailed to me by Elizabeth Slowley just prior to our interview-

"Neil, I have a little story to share; early this morning I was working in my garden. When I came by the avocado tree, I was thinking how nice it would be to have some avocados. The tree was trimmed back last fall however, so the branches are out of reach.

I went about my work and forgot about it. An hour later there was this loud crash and boom and the sound of things hitting the earth. Upon investigation, to my delight, on the ground was a branch broken off from the tree loaded down with avocados nearly ripe. What joy! I thanked the tree and now have plenty to share with friends at tomorrows' meeting.

What's interesting is how fast it happened and how powerful our thoughts are. Activating our higher brain power activates our whole body vibration.

Blessings, Elizabeth"

NS: "Tell me a little of your history..."

ES: "I stumbled on your site one day. I'm a massage therapist, but I've been in several accidents and I've had what I call 'Drain Bamage'"

NS: (laughs) "Was it actual brain damage?"

ES: "Yes, the whiplash and other injuries was actually the route that got me interested in massage therapy. I'm highly interested in brain function, and injury recovery. The upside to my injuries is that I became more compassionate and became interested in helping people to heal. I understand deeply what people are going through. I came across your site in the late 90's.

I understood the concept of stimulating the amygdala from your site, and from all of my other work it made perfect sense."

NS: "So when you began tickling your amygdala back then, how did you do it, and if you still do it, how do you do it these days?"

ES: "I basically just think it forward, I think of that whole frontal lobe area and make it come alive."

NS: "Do you see that part of your brain light up, or can you be more specific?"

ES: "Sometimes I hear a 'Pop!' I had one last week. One time it was so loud, it made me jump, and I actually looked around to see where it was coming from."

NS: "Is it a pleasant sensation or what?"

ES: "Most of the time it is."

NS: "How did you know that noise wasn't something else, outside yourself?"

ES: "Oh no, even that time when I looked around, I knew it was really me. I could identify the noise as an indicator of something internal changing in my brain, as others had also found to happen."

NS: "What would you say has been the general effects of incorporating that practice into your life? How has it changed things for you?"

ES: "Oh, absolutely. It's changed things with helping me really being aware and not falling back into a numb mind set. It helps me to stay alert."

NS: "So you feel like you're more conscious rather than reactive in your daily experience?"

ES: "Yes, I'd say so."

NS: "What if any effects do you think it's had on your health? You say you've had a lot of challenges. Explain if you can, the effects for better or for worse..."

ES: "I think that it's helped me be a lot stronger. I try to do it almost constantly. It comes up to my mind to 'tickle it forward'. I think of it all the time."

NS: "Do you think that helping other people to get well is important to keep your positive mind set and equilibrium?"

ES: "Oh yes, it's a very selfish thing to do, you know, helping other people. Because it all comes back to you." (laughs)

NS: "It sounds like you've had some problems, but you've risen above the inconveniences of any physical conditions…"

ES: "I've had to slow down, I've had to do things differently. Sometimes I still think I'm Super Woman, but I'm not. So you just have to take a different approach sometimes."

NS: "If you were to have any advice for someone who has never tried to tickle their amygdala, any words of advice or encouragement?"

ES: "I'd say keep trying, because it's well worth it- because when you're in that Reptilian Brain you can't think."

NS: "When you first started tickling your amygdala to get more energy into your frontal lobes, did you notice results right away or did it take a while to kick in?"

ES: "Actually right away, because I had already been pursuing that path, but without those exact words that, 'Oh yeah, that's my amygdala that I'm tickling.'"

NS: "Did it help to understand that there was actually a physical correlation within your head?"

ES: "Oh, absolutely, that's why I loved your web site because I'm very visual in that sense, to actually see it and think, 'Oh yeah, that's where it is, and this is what you do, and there it goes! It connected everything together, like the pieces of a puzzle. Absolutely. I use it a lot.

I don't have a car, so I take my bike or I ride the bus, and I encounter a lot of people on the street. Sometimes things happen on the bus, and things happen when you're out in public and you're there, so I'm tickling my amygdala forward and I'm beaming all this energy to people- 'Just calm down, how can I help?' and it's happened many times."

NS: "So you find you can do it any time or any place."

ES: "Absolutely. And you don't have to fall into the whole mob mentality of what's going on around you. It's like having something in your pocket, except that it's in your brain."

AMYGDALA TICKLE #18- "Positive Pulse Wave"
Visualize the positive energy you feel as a wave of energy emanating from your frontal lobes, going out and connecting with those around you- family, friends, and even strangers.

BIG PICTURE POSSIBILITIES

The big mistake most people make when trying to prove or disprove the existence and validity of Brain Radar is that they assume that such abilities can be or must come under one's conscious individual control and deliver results as if you are ordering from a Chinese restaurant menu, one item from column A and one item from column B.

This is an erroneous assumption that has little to do with verifying the existence of Brain Radar, and has everything to do with ego. Such a notion is obsolete and ignorant: The idea that if something is real, it must potentially be completely understood or controlled by your own individual self.

This is a left brain way of considering the You-Niverse. It assumes that the only thing that matters is under your conscious control and one inch in front of your face.

The need to exactly control Brain Radar to verify its existence is a left brain dictum. It is a tyrannical attitude that in order to prove that something exists, you have to be able to wholly define it and steer it in left brain terms.

The idea that you ultimately have total control over your universe, and that you can dictate exactly all that happens to you by your personal control is a fallacy popularized by a string of self-help books that purport to reveal an ancient "secret" formula, or the practical meaning behind quantum mechanics such as "If you believe something is true- it will come true!" or, "Just imagine what you want, and it will happen!"

Oh, that would be so grand!

Maybe not, actually.

Often, hell is getting what you thought you wanted.

But that's not reality, nor is it the lesson of quantum physics.

Quantum physics tells us *what is possible,* which is a great many things, no just the one slice of pie sitting on your refrigerator shelf.

Quantum physics also tells us that we, as observers, have an effect on what we are observing. This is a great deal different from saying, "If you imagine something, it will appear for you to grab!"

What quantum mechanics tells us truly, is that if we open ourselves to all of the possibilities that exist in the universe, then we have more choices- which is better than having our head stuck in the sand, believing in one and one thing only.

It's *not* that you can always materialize what you see in your Little Me Eye. If that were so, everyone would have three Ferraris in their garage and ten million dollars in the bank. And frankly, that's a totally naïve idea- there just isn't enough money in the world treasury, nor would the planet support so many gas guzzling air-spoiling cars on the road.

Everybody just can't have everything. (Nor would we want us to.)

The Big Magic I understands that you can get what you need, which is something else entirely, and by some accounts generally pretty close to what you've already got.

Real actualization does not mean that everything is accessible to you, only that some things that you may have not considered may also be accessible to you.

There is a HUGE difference between saying there are a great many things you may not have considered possible that are indeed possible- and that everything is possible to you at all times.

Aren't mental hospitals filled with people who are *absolutely totally convinced* they can fly, or something very much like that? Can they do it?

Brain Radar connects you to the real possibilities, even ones you have not thought of yet. It puts you in touch with what is real for you- not just what is pretend imagining.

Brain Radar works to ensure your survival, first and foremost.

In that way, Brain Radar doesn't care with the trivia of your own egotistical desires, your wishes for fame and fortune, or your need for some particular measure of success that we have culturally come to worship as a false god.

Brain Radar is not a card trick or an illusion you can call upon at will that will gain you notoriety on a television talk show.

It is a skill that is called into motion as you move in cooperation with The Big Picture, with the collective consciousness of what is truly good for you in the long term, and good for all.

Your brain is smart, and Brain Radar exists in the smartest part of your brain.

The sub-conscious and unconscious mind, the Big Magic I, is that grand storehouse of knowledge that extends beyond your immediate local environment. It is the World Wide Interbrain. And as such, you are only conscious of a little slice of it at any given moment, just like you can only look at one World Wide Web page on your computer screen at any given moment.

You really don't need to see the entire universe twenty-four hours a day. What would be the point? That's not a job for a human bean.

In this way, local consciousness is limited. It sees a small nice picture quite appropriate for a little guy like you.

But you are still connected to everything, have no doubt about it.

The Big Magic I can sense all of the quantum possibilities that exist beyond the reach of your nose hairs. When appropriate. When you need it.

That is the true nature of Brain Radar.

Bernd Jost
Senior Editor Rowohlt Verlag (Germany)

Among the hundreds of titles Bernd Jost has worked with in his long career, he selected The *Frontal Lobes Supercharge-* where the expression "Tickle Your Amygdala" was first coined in 1989- to be published and translated into German during his tenure as Senior Editor at Rowohlt Verlag in 2006.

Jost full well knew then as now that amygdala tickling does not require an individual to do any particular regimen, practice, or anything. You do it in your own way, in the way you choose, expressing the ultimate in independent democracy and self-determination.

NS: "Thanks for sending that picture of yourself- now I know what you look like after all of these years!"

BJ: "I thought you wondering, that's why I did it!" (laughs)

NS: "I'm keeping all of these interviews very casual, just like if we were sitting down to a cup of coffee. Do you have anything in mind that you want to say before I get to a couple of questions?"

BJ: "No, just start and we'll see where it's going, and I'm really interested in your new book…"

NS: "Okay, then. Can you tell me first about your education and occupation?"

BJ: "I went to high school and then to university to become a high school teacher, which I became for English and French language. I taught for a while and then went into publishing, but for a crazy company, so I went back to teaching for a year before I returned to publishing which for me was more interesting.

I started with non-fiction books and then I did everything, like classics, French literature, or funny books."

NS: "Did you always work for the same company?" [Rowohlt]

BJ: "No, I changed several times. I ended up being a crime fiction editor, mysteries and thrillers. I published a lot of American and English authors in Germany. Then my wife translated a book by Irina Tweedie who had gone to India for training by a Sufi master. She learned that there were other realities, not just this one you can touch.

At the time, neither of us really liked the book. I thought, 'I'm a typical European intellectual. I've read Freud, and metaphysical things are only for neurotic people. I'm not neurotic at all, so this doesn't apply to me.'

But then I found a leaflet that Lena was coming to Hamburg and I thought, 'Okay, let's go take a look at her.' And when I heard her, I fell in love with her. She was 73 at that time. She was a very funny woman, very much in life. I could feel that what she was talking about was real. So I started meditating.

She was completely without rules, and said 'You have to find your own way, there is no rule, no dogma, no nothing.' If there were some people that were proud that they were vegetarians, she ate meat! If someone didn't drink alcohol, she drank a lot of red wine! So she was someone who defied all of the rules, which I enjoyed. A lot of freedom.

So that's how I got into meditation, and I'm still doing it."

NS: "What kind of meditation do you do?"

BJ: "It's a meditation of the heart. You just go into the heart. If thoughts are coming you just drown the thoughts into feelings of love, because love is the strongest power in the universe."

NS: "So you concentrate on that heart chakra, as some call it?"

BJ: "You can do that, but you can just concentrate on the feeling. You remember someone you love. It might be your dog, it might be your partner. Or it might be God. The feeling, and the strength of the feeling is important."

NS: "You're familiar with that brain exercise in my previous book, The Mind Movie Re-Right. where one remembers and re-programs traumatic event memory into newly created positive scenarios. It sounds like you just skip to that last step in that exercise where the idea is to change a bad memory into a good memory where you see the event happening in an intelligent and loving way."

BJ: "Yes, in a way."

NS: "In regards to this new book, I'm thinking that the last bit of my book is going to ask the question, 'How do *you* 'tickle your amygdala?' implying that I can make some suggestions but ultimately each person is going to have to determine how they tickle their amygdala solely for themselves."

BJ: "Right. Of course. For everybody it's different."

NS: "So, back to your job, where did you eventually end up?"

BJ: "I ended up working for Rowohlt and I did that for nearly twenty-four years. After I started meditating I went up to my publishers and said, "Why don't we have a spiritual book series? I'd like to do that," and they said 'No way! We are a politically minded publishing house.'"

NS: "Tell me a little bit about that company..."

BJ: "It is a very big company, very intellectual, politically a little bit let's say, liberal. They publish a lot of American authors; they publish Jonathan Franzen, Toni Morrison, all the important American authors [back to Sinclair Lewis and Hemingway, through to Thomas Wolfe, Henry Miller, to present]. So it's quite a prestigious company. But they have to make money too, so they publish a lot of thrillers and entertainment books."

NS: "So you approached me about six years ago..."

BJ: "Probably seven. Yours was one of the last books I did.

I'm always on the lookout for interesting books, and in the end they agreed to let me do what they would consider more esoteric books. I was always interested in what was real, what is reality really like. And quantum mechanics tells us that reality is not what it appears to be, as Zen monks have been telling us for the last two-thousand years. So I tried to find books that crossed this gap.

I published this book by Michael Harner on shamanism. I published a book by Gary Zukav called *The Dancing Wu Li Masters*, a book that combines science and spirituality. And a lot of those kinds of books, and they were quite successful. But in the end, the management changed, then changed again, and the new management didn't want that kind of book so they stopped it, and I was left with my thrillers and mysteries. *But*, they then started with the new

"How To Do" kind of books and I was in again, and I thought 'Brain is an interesting subject,' and I found your book and I really liked it, or I wouldn't have done it!" (laughs)

NS: "So what was your impression of this idea of tickling your amygdala that I first introduced in *The Frontal Lobes Supercharge*?"

BJ: "Well I was already familiar with those kinds of results from my mediation, and it seemed to be a shortcut for people."

NS: "There are a lot of self-help books out there these days."

BJ: "Most of them are just rehash. (laughs) I'm quite interested in new therapies and consciousness type technology.

There's those books that promise that all you have to do is just obey the law of attraction, and then you get the big car and the big house and lots of money!" (laughs)

NS: "Oh yeah, of course. This is the fake myth that we see in a lot of those books. I'm just trying to point out that if you try to solve everything with your Reptile and reactive brain, you're sunk. You've got two toolboxes in your brain, one in your frontal lobes, and it's got all kinds of tools in it. But then there's the toolbox in your reptile brain- and it has nothing but a hammer in it… or a club." (laughs)

BJ: "Probably a club."

NS: "Well, sometimes you need a club. That's fine if you're trying to hammer a nail into a wall- but you can't build a house with just a club. But that's what people try to do. And if you're clicked backwards into your reptile brain, that's the only tool that you've got."

BJ: "Well all the movies we are shown, they all feature the club. Even the movie like *Avatar*. In the end it's just brute force, which is ridiculous. That's the message of the film."

NS: "When you look back on your life, what are your successful strategies? What have you used to guide you?"

BJ: "I've relied on people and some very good friends that I treasure. And for me, music is very important. I put on my headphones and I'm off for an hour, and when I come back I'm happy again."

NS: "Why do you think that is?"

BJ: "Music, in a way is structured, and it reaches your emotions."

NS: "Does it provide you with an emotional mental balance that you wouldn't attain otherwise?"

BJ: "Actually, I've read a lot of books about it, but I'm really a soul fan. I love black soul singers. Aretha Franklin, for one, or Etta James, Ester Phillips. They sing from the heart. I think you underestimate the heart. The power of the heart is even stronger than the power of the brain."

NS: "I think people misunderstand me, and I understand why.

When people speak of the heart, they're speaking of empathy, they're speaking of compassion. These are legitimate functions of the brain. I think the evidence for that lies in examining those who have had damage to their frontal lobes [medial pre-frontal cortex]. Then the ability to feel what someone else feels disappear. That ability to be in someone else's shoes is a frontal lobes process."

BJ: "For me, meditation has definitely been a key tool. I've meditated so much that most of the time my amygdala is clicked forward. It's very rare that I'm really angry and clicked backward into my reptile brain.

I've played around with several methods, but I've always come back to the heart meditation."

NS: "Why do you like that method?"

BJ: "Because I sit down and after five minutes I'm off- I'm not there any more. And when I come back I don't notice any noise, I'm really gone in a way! I'm relaxed, I'm in harmony."

NS: "At times when you do feel agitated, and you sit down and do this meditation, does it relieve the agitation?"

BJ: "Oh definitely. I'm not getting agitated so much as I used to get. You know, publishing is a very hectic job that always has disaster happening and you can get caught up in it. Through meditation I learned to take a step backward and just take a look at it and say, 'Well, you can do this or you can do that." I became much better than I was in solving crisis and problems because I wasn't caught up in the whirl."

NS: "I think the heart meditation you describe is valuable because people instantly understand that iconography. The idea of the heart is a symbol and archetype that people have used for millennia. So people get that instantly, and it's absolutely valid and useful. You get no argument whatsoever from me.

But I don't think you would be able to do that meditation if you didn't have a brain."

BJ: "Definitely not!"(laughs)

NS: "I'm all for any type of meditation. It's already been proven, the benefits of a myriad of forms of meditation. But conversely, I think there are people who get by fine without it."

BJ: "Oh definitely. People get immersed in something like painting, or making music, or going hiking and being in touch with nature. It's all the same! It's meditating, or tickling the amygdala forward. It doesn't matter."(laughs)

NS: "When people think of meditation, they think of sitting in a lotus position or staring at a candle for an hour. But like you said, everybody has their own way."

BJ: "For me, it came rather late because I'm a book person. Books did it for me, and it took quite some time. When I went to California and I was sitting at some kind of lagoon and suddenly I and nature were one. I had read about it, I knew about it intellectually- but suddenly I experienced it. And it changed a lot for me. Now I can meditate in nature by admiring a flower or watching birds, or something like that."

But it all comes back to what works for you."

NS: "You just have to listen to what your amygdala is telling you…"

BJ: "It's the same with food- I don't have to read all those medical books that tell me what food to eat and what to not, I just listen to my body. Sometimes it likes chocolate and sometimes it likes rice and beans." (laughs)

NS: "Listen to the internal signals. There's lots of good information out there, but..."

BJ: "I had a Twinkie once because I wanted to have the experience- but never again!"

NS: "Maybe that's the lesson in our discussion here, listen to what your body and brain are telling you, and to hell with the experts..."

BJ: "To hell with the experts!" (laughs)

AMYGDALA #19- "Heart Wave Energy"

Sit for a short time and remember the feeling of love you have that you associate with anyone- a family member, a pet, a friend, or even a stranger.

Feel how you feel when someone shares that energy with you.

See your entire being, body, and brain light up with a warm glow.

Then, project that heart energy outward, and share it with another-a family member, a pet, a friend, a stranger- and perhaps even most importantly- someone you might mistake as "an enemy" in conflict.

BRAIN RADAR EXPERIMENTS:
FAILURE AND SUCCESS-
LOTTO AND RAINMAKING

In my early days of exploring the potential of the human brain thirty years ago, I was, like many, drawn to the idea that our brain has infinite potential. I was captivated by the idea that my brain was a computer of enormous capacity beyond my wildest dreams.

This attitude, in itself perfectly correct, was fueled by all of the many examples I had seen of super-human abilities performed by the very same sources I've quoted a few pages earlier in this book and the dozens and dozens of laboratory experiments that demonstrated the real existence of ESP and other paranormal abilities.

At some point I decided that there would be no better way than to focus my own latent powers than in predicting and buying the next big winning Lotto jackpot ticket.

So I set the task upon myself to meditate, think positively, to visualize, and do whatever else I could conceive of to be the next big winner in and upcoming Big Super Duper Multi-Million Dollar Drawing. $$$$$ Whooopeee!

After all, it seemed quite obvious that seeing into the future, and in fact manipulating my own very future was indeed possible and clearly demonstrated by all manner of persons. Why not think myself into becoming a big, rich, lucky person?

Certainly, there was no cosmic or other law that said I should not be a gazillionaire, and everything around me said that I would be a gazillion times happier if I was rich.

All indications was that the human brain could tap into the vast super-conscious mind, and side-step a lifetime of hard work at jobs I might not be crazy about, and I could instead turn on my Super Brain ESP and pick the lucky winning numbers.

So, I set about my task, to turn on the infinite powers of my frontal lobes to predict or manipulate the Colorado State Lottery.

I did many things to this end:

I sat in the Lotus position, with my legs crossed, eyes closed, and meditated on perceiving the winning numbers.

I instructed myself to dream the winning numbers each night before bed, and kept a handy pencil and paper on my night stand so that I could quickly write down when these digits projected themselves into my dreams.

I visualized myself receiving the giant six-foot wide check, smiling for the camera after being presented with my gargantuan twenty-million dollar winning prize.

I carefully painted my winning lotto mandala, with the winning numbers as six points on a cosmic pentagon around a cosmic pentagram.

I even went to the TV station where the drawing took place, and secretly buried my chosen numbered ping pong balls at each corner of the building, and at the front and back of the station to influence the drawing.

And twice each week, I bought my winning lotto ticket and faithfully awaited the announcement of my winning lotto drawing numbers.

I waited for the cows to come home from Pluto.

My lotto experiment was a complete failure. I had no more success in picking the winning numbers than I had success in flying into the top branches of the tree in front of my house by flapping my arms up and down.

So it goes.

Yet, I did not completely abandon the idea of tapping into the great Universal Energy System. In fact, as the years went by, I studied how the human brain worked even more intently than ever. I was determined to understand how my brain worked. I was determined to discover what the working rules of real Brain Radar really were.

A couple of decades later, I found myself as a guest on Coast To Coast AM, the world's biggest night time radio talk show, hosted at that time by Art Bell.

I had quite a number of interesting if not outright remarkable stories of real paranormal experience to tell, as well as more down to earth and scientific explanations of how the human brain worked to relate. As Art's guest, I spoke of easy to learn methods, including amygdala tickling, that would result in remarkable and measurable increases of creativity, problem solving, intelligence, and automatic reward and pleasure stimulation inside one's own brain. And it could be done for free.

After much time, I had realized that it didn't matter any more whether or not one had a million dollars in the bank. In fact, one was probably better off without the burden of great and enviable material wealth.

One of the noteworthy early experiments we did on Art's program was the first mass audience brain focus experiment ever conducted anywhere. It was a thought experiment of not just a group of people in a room or even a large auditorium- It was an experiment that I had designed that involved Art's complete radio audience: Millions of people.

The object of the mass brain focus was to put out a fire. But not just any fire. It was a big fire that was occurring throughout the state of Florida at the time. This is what happened:

The nation had been reading the headlines for several weeks in June and into July, watching with horror as the state of Florida had been experiencing a disastrous drought with record high, brow-melting temperatures and widely spread uncontrolled wild fires devouring homes and property unabated.

The worst damage was occurring in northeastern Florida. From the Daytona Beach News Journal:

"...It is not likely that residents of Flagler County will soon forget the 1998 Fourth of July weekend. Discussions about fireworks and festivities became a moot point Friday morning as the entire county was evacuated in the fear that the most populated areas of the county would be consumed by flames. Mass hysteria settled down to grim resignation Friday as many slowly made their way out of Flagler County in bumper-to-bumper traffic..."

Before going back on Coast to Coast I had informed the producers that I would try something unprecedented- A mass brain focus experiment involving the entire radio audience in an attempt to cause a substantial change in the weather in the most threatened area of Florida, in the northeastern corner of the state.

I had read in the newspaper that although people had hoped for some relief from rain, little was predicted, certainly not enough to quench the flames that were raging across that region. The fire fighters said that it would take something on the order of ten inches of rain to have any impact on the fires. That much was in no way anticipated by the weather service. (The extended forecast for that region predicted only a 40 percent chance of rain for the region, July 7, *Daytona Beach News Journal*.)

A quick glance at the satellite picture on July 7, the night of my show, showed no clouds at all over all of Florida nor for hundreds of miles in any direction. There were significant dry areas without any precipitation throughout the entire country. The notion that a bank of clouds would descend upon this part of the Florida and remain hovered over the target area and stay there to pour something in the order of ten inches of rain was something that would be impossible to foresee and predict, and no one had done that.

But that is exactly what I proposed to do.

I went on Art's show and instructed the audience on amygdala tickling and focusing their brain energy forward to deliberately and consciously increase frontal lobes C.I.C.I.L. activity in their collective intelligence brains.

Then I went out on a limb.

I suggested that if we could combine our millions of brains into one advanced Collective Intelligence, that we could affect the weather and cause rain to occur in the northeastern part of Florida, enough rain to help put out the fires.

Although though such a thing had never been attempted before on such a scale, involving people scattered across an entire continent, I said that it would likely take a few days to manifest. Even a bean takes a few days to sprout.

I then proceeded to lead this mass radio audience in our brain focus Brain Radar experiment, to bring relief to the burning residents, animals, and flora of Florida.

That night, even while on the air, I began saving weather satellite images and Doppler radar images of the target area for documentation. (See: http://www.neilslade.com/weather.html)

And we all waited to see what would happen.

Before the show was even over, clouds began to appear in the satellite picture where none had existed before.

Within twenty-four hours, cloud cover and some heavy rain in areas of Northeastern Florida. By Saturday, the heavy rain began.

And it rained.

And rained.

And rained.

Days later I began to receive emails from Florida:

"Neil! Shut off the tap! Enough!"

After the collective brain focus amygdala tickling, the immediate changes in Florida can unambiguously be seen in the weather record for the state:

Despite record high temperatures and excessive dry weather in Melbourne, Orlando, and Daytona counties June 30, July 1, 2, 3, 6, 8, and 9, the heat wave abruptly ended after July 9 and for the remainder of the month. The immediately preceding drought was replaced by excessive rain in the time period immediately following, with 262% greater rainfall in Melbourne, 146% greater than normal in Orlando, and 131% greater rainfall in Daytona.

(See http://www.srh.noaa.gov/mlb/?n=summer1998)

It wasn't Noah's flood, but it was sufficient to end the wildfires.

Brain Radar had made its indelible mark, and the entire country had been witness.

Coincidence?

I suppose Life On Earth is one big coincidence if you want to see it as such.

You generally see what you want to see, and what you expect to see. When you are free of expectations, you see what IS.

George Noory
Syndicated Radio Talk Show Host "Coast To Coast", International Broadcaster

George Noory has for the past decade been the host of arguably the most popular night time syndicated radio talk show in the world, *Coast To Coast AM*. His show regularly and seriously examines charged and controversial topics often neglected by other broadcasters such as strange phenomenon and paranormal activity.

His openness to new ideas, his eagerness to learn more, and his overtly friendly manner on the air makes the show a favorite among millions of regular listeners every night.

NS: "You're in a very unique position, and that's one reason why I wanted to talk to you for this book. You've talked to so many different kinds of people and you've done it for so long. But I first wanted to ask you something about yourself- How did you first know what you wanted to do in life?

GN: "That's a good question. These things start often when we're very young. I had an out of body experience when I was eleven. And my mother was very influential, although she was very religious.

The first book she brought me was *We Are Not Alone* by Walter Sullivan who used to be the New York Times writer. I was just immersed in that from a very young age. I just decided that I wanted to unravel these stories. I tried to say to myself at that age, 'What can I get into to allow me to try and cover these things and find out what's out there?'

And it kept coming up that, 'You've got to be a broadcaster,' because these people will then talk to you, the Neil Slades and the Stanton Friedmans. I thought, 'If you're a real broadcaster, then you can interview them!' – But, you're not going to do it if you just call them up and say, 'Hi, I'm George Noory and I'm fascinated with this...'

So I decided at a very young age that this is what I wanted to go into in order to cover all of these various subjects that I cover today."

NS: "Tell me a little bit about that out of body experience that you had, because it sounds like you were motivated by something that wasn't a rational event or a rational decision, but rather by something that came from this other part of your mind or brain that defied logic."

GN: "Well, whatever happened to me, just happened. I woke up and I thought I was dreaming because I was looking down at my body. I had stayed home from school that day, I had a fever."

NS: "How old were you?"

GN: "I was eleven years old. So I'm looking down at my body and I'm going, 'My gosh! What is this!?'

In that instant when I realized that there was something strange here, it rammed me back into my body and I woke up in my physical state. I knew something strange had happened to me, although I didn't know what it was.

So I went to the library- we didn't have computers then- and I stumbled into this book called *Projection of The Astral Body* by Sylvan Muldoon and Hereward Carrington. And I read it and realized that's exactly what happened to me.

But it was a weird experience- I was bouncing on the ceiling. It lasted just a few seconds, and then down into my body I slammed, and that was it."

NS: "I've had that exact same thing happen to me, so it's very interesting to hear your story. It was distinctly not a dream, or my imagination, but something altogether unique. I wasn't sick either.

I was older, about fifteen, and I was home from school and got into bed to take a nap, and the next thing I knew I was up on the ceiling and I realized, 'Oh! That's me down there!' and then Boom! It was so shocking that it was, Whoooshhh! right back into the body. But I'll never forget it.

Have you talked to other people who have had that same kind of experience?"

GN: "Oh yeah, I've interviewed people like Robert Bruce, who is an expert in this. I never had an opportunity to interview Robert Monroe, who is gone now, but who was of course a giant in the field of astral projection. But everybody agrees, I had an OBE.

It's interesting that I haven't been able to duplicate it, although other things have happened to me since then. I've become extremely intuitive, strong in my dream state. So the power of the mind for me has really been multiplied tremendously even though I can't voluntarily do an OBE."

NS: "What other things had you tried other than being a broadcaster that weren't fulfilling or satisfying? How were you certain that this was the right field for you to be in, no matter what?"

GN: "I had not at that age considered anything in the way of a career until that event. But after that happened at the age of eleven, that's all I thought about. That's all I did.

I mean, I read books on Edgar Cayce, I joined NICAP which is the National Investigations Committee On Aerial Phenomenon- it's what MUFON (Mutual UFO Network) is now. I knew that's what I wanted to do.

When I got older in life, my dad was paying for my college and wanted me to become a dentist. I didn't want to do that, but to appease him I did pre-Dental for two years at the University of Detroit.

And then one day by chance- you talk about fate- a friend of mine who was studying broadcasting there and who was working at a TV station said to me, 'Hey, I want to go home for Christmas'- he lived in Cleveland at the time- he said, 'They won't let me go unless I can find someone to replace me.' He was a copy boy making two-bucks an hour. So I said, "Yeah! I'd love to do that.' And that was when I had my first real experience in broadcasting.

I realized, when I was eleven, this was exactly what I wanted to do. And now this just convinces me, so in a strange way that is what got me back on track. I switched my major to broadcasting, and did not tell my father about it.

My dad got my report card, because it was always mailed to him. He said to me, 'George, I've got some good news and some bad news.' I said, 'What?'

He said, 'The good news is, whoever this guy is, he's doing great in all these communications courses Speech, Journalism, Radio Production. He's getting A's in everything! The bad news is- where's your grades?'

NS: (laughs)

GN: "So I said, "Dad, I switched my major.'

And he didn't talk to me for two months. But one day he saw my name show up on a television screen as a credit at the end of a newscast that was broadcast worldwide, and he said, 'Oh my gosh! Is that *you*?'

And I told him, 'Dad, I told you that's what I'm doing now in school- and everything for him changed too because he began to realize that maybe all this wasn't that bad.

And that started my career, and that's what pushed me in this direction. I always, I always wanted to cover these strange stories."

NS: "That's a really interesting story about how you got to where you are. I think people are very much affected by forces outside themselves that say, 'You should do this, or, you should do that.'

But it seems as though the people who are the happiest, in spite of that, have an internal compass as it were. And if they follow that compass it doesn't matter what else is going on around them, that they have a sense that '*this is the right thing*'. It sounded like you made that connection, despite what others were saying about what you should be doing."

GN: "Absolutely. You have to go with that inner feeling, that drive within you. You have to try, and I did.

You know that Robert Frost poem- *The Road Not Taken*? I truly took the path that was not traveled. And for me it has made a world of difference. It's been a life changing decision for me that I made a long time ago, and I never regretted one moment of it, ever."

NS: "You've talked to so many people over the years at your job. Everyone is looking for happiness and fulfillment, and you've been in a position to observe so many people doing this, that, and the other thing. Can you pinpoint something- in your opinion, what is the biggest mistake people make, if you could nail it down to one thing?"

GN: "We're all human, and we all experience the ups and downs of life. The secret is to rebound a bit faster when the things are down. One of the things that I've been able to do is to try and erase some of the negativity that there is out there, and to keep forging ahead.

I have never been truly successful prior to Coast To Coast. I mean, I made a nice middle class living, I lived in the mid-west, I had my own little talk show- But there was a time in my life when everything collapsed, I got hit over the head financially, I lost my shirt on a number of little projects that I did, things weren't going well, my health was going down the wrong way.

But I stopped feeling sorry for myself, I stopped feeling negative about things, and I pulled myself out of it, and I never stopped after that. I remembered a lesson that I had always known, and that was to believe in yourself and never doubt it.

Trust your instincts, use the things you've learned in your life, and just keep pushing and forging ahead. That's what I've been doing over the last eleven or twelve years now, and it's made a world of difference for me."

NS: "It seems like we're often ruled by our emotions. A mistake that we make is that we make emotions the captain of our ship. It sounds like as though you're saying that it's not the captain, that there's something else. We can rule our emotions, or at least not be driven by our emotions. There's something above that which ultimately gives us control over how we feel in life."

GN: "I've learned that if you don't push for things in life, but merely get ready for it, prepare yourself for it- and for me, the example would be Coast To Coast. I wanted to do it, I thought about doing it, I dreamt about it- and it happened. I was prepared for it when it happened.

I had studied it, I understood the genre, I knew many of the guests that Art Bell had already interviewed, so I was ready to step right in, in a moment and not miss a beat.

My daughter had been applying for some jobs and had been calling people back. I said to her, 'Wendy, don't call them back. If they want you, if they're interested in you- they'll find you, believe me.' She's starting to get that now, too, that pushing and pushing and pushing doesn't always get things done. Put it out into the universe then let it go."

NS: "I noticed that there are two different kinds of people- those who say, 'Just be, just let things happen...', but that doesn't seem to work out very well. Then there's another kind of person who tries to plan every single thing, and control every single thing. That doesn't seem to work out well either, because those people can't seem to enjoy what they have at the moment.

However, there seems to be this balance where you can be in the moment but you're also simultaneously planning for the future. In that balance between the two, that's when things really work."

GN: "Absolutely. There's no question about it. It's worked for me time and time again, Neil. That's what I try to teach people on Coast To Coast- 'Be ready, but don't push it, and let it happen.' Whatever is out there in the universe, let it do what it does..."

NS: "But be prepared, prepare yourself as well..."

GN: "Definitely- that's critical."

AMYGDALA TICKLE #20- "Prepare A Left Frontal Lobes Recipe"

Be prepared. Start with a Recipe List of Ingredients. Make a list and WRITE IT ALL DOWN ON PAPER. Stick it up on your refrigerator to remind you till you get it made.

1. Decide what you want to cook up.

2. Decide what you need to prepare and have in place.

3. Go through the list, one item at a time and obtain and prepare.

4. Put it in the oven, and like a cake baking- let it do it's thing and don't keep poking it every five minutes to see if its done!

Palma Lee Stephens
Neurosurgery and Medical Transcriptionist

NS: "Let's set the tape recorder levels here... say something..."

PL: "I enjoy watching pictures of Erfie and Chloe. That kind of makes my day..." (laughs)

NS: "Perfect. That works. (laughs) Okay, so when did you first learn about tickling your amygdala?"

PL: "It's been more than ten years; it was in the late nineties. I've always been interested in the brain and controlling your thoughts, and visualization. Clicking the amygdala was right up my alley when I saw your site."

NS: "Well, the difference between what I write about and other behavioral modification things is that I'm actually talking about physical structures in the brain and what they do."

PL: "I'm a medical transcriptionist and I worked in a neurosurgeons office. So I'm very interested in the physical aspect of brain function."

NS: "When you first started looking at my web site, with your background, was there anything that raised a red flag?"

PL: "No, no. I knew that you knew what you were talking about. It all rang true for me."

NS: "Where did you go to school?"

PL: "The University of Florida."

NS: "Tell me a little bit about your own experience- what kind of experiences have you had relating to controlling your brain? What have you employed over the past ten years that you've found effective?"

PL: "I use it in just every day small ways- I might get angry with other drivers when I'm in the car, so I do the amygdala tickling there and it always makes me feel calm and peaceful and in a pretty good mood. I use it practically every day."

NS: "A lot of people just read about amygdala tickling and about frontal lobes and such, and then they don't do anything more with it. But you actually use that technique. And you work in a neurosurgeons office. Gee whiz. What a shock!" (laughs)

PL: "I think that people wait until a big trauma comes along, but then it's less likely to work. If you use it with every day small things, then you automatically do it, instead of waiting until something horrible comes along. Then life is easier to handle."

NS: "How do you do it?"

PL: "I don't always use a feather. When I sit down and I'm quiet and focused, then sometimes I'll see the feather. But when I'm busy and I'm distracted with other things, perhaps when someone is talking to me- If I'm in an environment that's chaotic I sometimes forget about the feather and I just see it click forward. I just see the amygdala on both sides of my head, and it clicks and sends energy forward. It just takes half a second."

NS: "What goes on inside your head once you've done that?"

PL: "If I'm agitated it centers me and I feel calm and things feel more orderly around me. That's the feeling that I get from it."

NS: "It sounds like you make these small micro-adjustments through your day that keeps you moving in the direction that you want to go by tickling your amygdala forward."

PL: "Right. I don't have these big hard moments that come up, I think life has been easier since I've been doing it. I don't have these big traumas, or big dramas. When I was younger I used to have these big dramas, but I have a lot less of that.

Things happen and you can fall into despair momentarily, but when I do it, the tickling, you find that hope that things will get better. And things always do get better. It puts you in the right place mentally."

NS: "It does actually change the way in which your brain works so you can solve whatever problems come along with the higher functions of your brain."

PL: "For years when I was younger I had old anger from past family issues, and I couldn't get rid of them. That doesn't exist anymore."

NS: "I've observed in people that this is not a universal given, that as you get older you get rid of these old issues or issues of anger."

PL: "No it's not. My father, until his dying day, had things they he kept going back over, there were still things left over, things that had happened to him as a child."

NS: "Is there a spiritual component of your amygdala tickling?"

PL: "Well, I do some meditation, and this Flow Dreaming thing, and I combine the amygdala thing with all of that. It's all going in the same direction. But the amygdala tickling is very physical, it's anatomical."

NS: "People do a variety of practices- physical exercise, meditation, yoga, or aerobics, people have their religious faith- so there are all of these things that we can incorporate into our daily lives that reflect frontal lobes activity. But you've also chosen to integrate this conscious amygdala tickling with other things you do- why bother?"

PL: "Because I like the brain. I can see that part of my brain any time, when I'm walking around. I love the brain. I use it in a different setting than I would the other things. I'm sort of 'omni', and I use things in different ways.

I like metaphysical things like talking to angels and such, but I like tickling my amygdala because it's real, it's physical, and because I'm doing something with my brain.

Tickling my amygdala is in the three-dimensional world. I believe that from the world of ideas we create our physical world, and I believe in an invisible world and a spirit world. But amygdala tickling is almost like taking a physical pill. I understand what the frontal lobes do, what cooperation is, and imagination. It's more believable for the three-dimensional mind."

NS: "I see it as having a certainty. We know what the frontal lobes do. We know we have emotions. We have all of these tools in our brain, and perhaps the reptile brain is just a hammer we need on rare occasion. But why limit yourself to just that tool?

Hey, I think we just invented a new metaphor right there. (laughs) There's a bit of cooperative creativity right there- Together trying to formulate this thing, we invented something, 50% Palma Lee and 50% Neil Slade. [And so we did in this conversation.]

The reptile brain is sometimes like a little hammer, sometimes it's like a big sledge hammer. You just pound on something and that's it. But it is certainly no Swiss precision eyeglass screwdriver. It's not a band saw, it's not a tape measure, it's just a hammer- bang! And you *can't* build a house with just a hammer. So thank you very much!"

PL: (laughs) "Sure! You're just bouncing it off me!"

NS: "So there's a certainty in tickling your amygdala and turning on the frontal lobes in the same way that there's a certainty of putting a shoe on your foot that allows you to walk comfortably over pebbles."

PL: "Yes. It's more down to earth. There are other practices that I do, which I enjoy, but I do what is most comfortable depending on where I am. I understand what the frontal lobes do. During my everyday things, I do the amygdala thing, and it's had a major impact."

NS: "Well, you've been a great inspiration to me over the years, thank you very much."

AMYGDALA TICKLE #21- "Hammer Versus Tickle"

Get
    a hammer and get a feather and another person.

Have this lie down on the floor, eyes closed. The object is to get them to laugh or smile, and not to take you to court.

First apply the hammer.

Then try the feather.

THE MORAL OF THAT STORY

Just in case you missed it, Brain Radar fails when you are just thinking about your own little world and yourself. That's why the list of winning lotto ticket holders is not surprisingly completely absent of psychics and all those people who "Really truly" affirm and desire to be big millionaire winners. No "The Secret" there.

When it's just all about Me Me Me, the universe typically says No No No.

In the case when you are praying for somebody else, when your focus is on the good will and the survival beyond your own underpants, then you have a good shot at success with Brain Radar magically guiding you all the way home. It's that simple.

The team– the rest of the universe that sits outside your front door– supports the member that works for the team.

The member that does not work for the team is expendable.

What team are you on?

Brain Radar is perception of The Whole, The Big Picture.

If you are using only your reptile brain, that computes nothing but Me Me Me consciousness, then you have an Ego Blind Spot

You only see what is just in front of your nose, a little nearsighted sliver.

Debra Ann Robinson
Director Himalayan Children's Fund, Meditation Instructor

NS: "Tell me about The Himalayan Children's Fund."

DAR: "This is a U.S. non-profit set up for the purpose of helping the poor and impoverished people in the Himalayan region by Thrangu Rinpoche, who fled Tibet at the same time as the Dali Lama in 1959. He is a scholar and a master, a high lama in the Kagyu lineage of Tibetan Buddhism. I became Director of this project fund in 1989."

NS: "Of the people I know, you've seen the real extremes of living conditions. I haven't been to India, I haven't been to the mountains of Nepal. And I expect many of the people who read this book will be fairly well off compared to the living standards in many parts of those countries. I think there is a real confusion, especially in this country and other places where we have so much *stuff*, that we confuse happiness and the fulfillment with life- we mix it up with the exterior world."

DAR: "But that's also been true forever, all over the world. I was trekking in the mountains of Nepal and there was this kid high up in the mountains wearing a Michael Jackson glove. Remember in the movie *The Gods Must Be Crazy* when a tribe of remote African Bushmen find an empty Coke bottle and they all start fighting over it?

Human beings have this 'Oh, what's going to make me happy, and I want it!' thing. Look through history, it's always there, 'I want what you have!' and we're always willing to clobber each other trying to get it."

NS: "But personally you've seen the extremes- you've been to the ritzy country clubs here, and hung out with people who live in these huge mansions who are enormously wealthy. And then you walk through India, and as you told me earlier- you see a whole family living in a tent on the sidewalk the size of my coffee table, and they're laughing and making jokes with virtually not much more than the shirts on their backs."

DAR: "And I've walked through extreme wealth in India and extreme wealth in Nepal as well too."

NS: "So, can you make a blanket statement that one's happiness and contentment is not dependent upon exterior circumstances, and that it's more about our own inner perspective and how we relate to the world?"

DAR: "Exactly. Yes, it is so important that we begin to recognize our interior life and how we're working with our mind in relation to exterior surroundings and begin to see that the exterior doesn't really have any bearing on mental ease and clarity. That's what personal liberation is really about.

Even just your physical body- is that *you*? Who's what? What is it? The looking into your own mind to realize the nature of your own mind, the nature of all things- the *essence*- not just all the *appearance* of things. But we've been looking the other way a long time and it's so strong...

It's all like a dream, it's all coming and going- it's all temporary.

So, in this temporary thing- what is real? In the end, we finally get to the path of compassion and non-harming. That's what it's all about, to be concerned with the suffering of others, and to help relieve it."

Jim Casart
Financial Consultant, Certified Public Accountant

My friendship with Jim began long ago when I began to give piano lessons to his two daughters, and guitar and piano lessons to himself. After his children grew up and went off to college and careers, we continued to play music together on a weekly basis, which we still do to this day after more than twenty years. He has been a financial and tax consultant to many large businesses and individuals over a life time of work. He lives a professional life quite different from my own, and his experience and perspective is something that I have always found very valuable in many ways.

NS: "I'm seeing a thread that ties everyone's experience together, based on fundamental brain physiology that shows that everyone relies on their amygdala. It doesn't matter whether you're a CPA, an attorney, a garbage collector, or a concert violinist, or whatever it is you do- we're all working off that same circuitry to provide us with feedback and guidance.

So, can you illustrate, as a guy who works on people's finances, where you've seen people react backwards into this very primitive, defensive fight or flight posture and how that's worked to their detriment, or perhaps conversely where people have tickled forward into their frontal lobes and it's been to their advantage?

JC: "I believe that some people might have a predisposition towards optimism, which might come from birth but is probably affected by their experiences in life.

For example, some people with unbelievable disadvantageous conditions that you and I might call absolute miserable poverty, have just looked around and said to themselves, 'I have what I need, and if I don't, I can get it,' whether they're in Bangladesh or New York City.

That seems to be a very, very basic factor in human happiness. You simply believe there's enough for you too. So, that means you don't have to go to war to get it to take it from somebody else. Whatever it is that you need is there and it's available. Often, 'give to get' is a good representation of that-

NS: "Give to get?"

JC: "The more you give, you just assume that's a good thing, and that the universe is abundant and you'll get what you need if you give what others need. I think, even if you're fooling yourself a little bit, this seems to be a much healthier way to exist than to assume the opposite.

The opposite way of thinking is to think that there's a scarcity of good things and to take that one step further, if 'I want something and if somebody else has it, I just take it from them.' That kind of approach to life doesn't exist with an 'abundance mentality', the idea that you can cooperate with someone else to multiply good things."

NS: "So when you talk about abundance, are you talking about a reality or just a philosophical point of view?"

JC: "I think the philosophical point of view when taken to its natural conclusion will blend into the reality, it becomes reality. Two people coming together working towards something in a cooperative fashion will change the reality. You can affect things by what you think about it, your observation affects what's around it."

NS: "In your work you've been around people with tremendous wealth, and you've been around corporations that had tremendous resources- and you also play music with me, so you also know just the opposite of that- (laughs) So, can you talk about this in real terms?"

JC: "We've all seen famous examples of people who seem to have all that they would want, and people reach different conclusions about that. I've seen people who have made the choice to give away roughly half of their wealth for the betterment of the world. That seems to me to be personally healthy.

One of the philosophies you've probably run into is the idea that every loving act is selfish. What better gift can you give to yourself than to love something- you always get more than what you're giving. If you bring that into the financial world, there's a discernable difference- not only that, it hits you right in the face.

My observation among a significant number of wealthy people is that the ones who have a concept of 'enough' are truly blessed. Wealth isn't blessing them, because 'enough' is in the eye of the beholder. And yet those who don't seem to have enough, and they can be extremely wealthy, are cursed. That blessing or curse is something that occurs over a wide, wide range of nominal wealth, and can be something that one can choose to have or not have."

NS: "What you're saying is that it doesn't matter how much money you have in the bank, your observations in the financial world lead you to the conclusion that it's a matter of whether you adopt this philosophy of 'enough'."

JC: "You might call it a philosophy. If you look at twenty million dollars, most of us would say that it's 'enough'. But I've see a lot of people who would look at twenty million dollars and not even begin to understand the concept of 'enough', and it's kind of scary.

I have great admiration for all of my clients, and feel personal empathy for them. The ones who have that gift of knowing when enough is enough- and that is probably well short of twenty million dollars- then they can find other things in life that are more personally satisfying than chasing their twenty-first million."

NS: "I think what you're talking about in terms of wealth could be applied to anything, be it a romantic or marital relationship, or health, or material wealth, or political situation- we always look to improve things- but the glass is either half full or half empty."

JC: "I like to think that it's a choice we can make, because there are very, very few things in this life we can control outside of our physical reach. But it certainly includes the stuff inside our head.

What a blessing, if the things you can control are *right there* in constant contact with you. If that's where the switch can be flipped to turn- not to change the reality, but to change the perception of reality... What a great place to start, to get up in the morning and to stare in the mirror and say, 'The answer to my problems is staring right back at me.' This is a very empowering thought.

On the other hand, if you look in the mirror and see the enemy, and think "Whoa! That person is going to dog me all day long...'

If it's a matter of choice, I know what I'd choose."

NS: "We've previously talked about brain function, and the manner of reacting in a non-thinking manner, in an aggressive fight or flight, attack-counter attack manner. But people also have this capacity to process things in a cerebral way, that's not devoid of emotion, but in a way that includes cooperative, creative, higher intuition and logic..."

JC: "The function of the amygdala to allow us to react in a quick and efficient automatic way to avoid being swept off our feet by an attacking pterodactyl has proven itself to be useful thing in the past. But most things aren't life or death situations. That momentary extra processing time to use our frontal lobes is probably well worth it. That's where the richness of our life is."

NS: "I agree. The whole idea behind tickling your amygdala is so you just don't react out of fear, and instead increase the amount of frontal lobes processing that you do to address problems, and increase creativity to get more enjoyment out of life. But we can easily fall into the trap of coming to rely on that system of automatic reactions."

JC: "One of the joys in life is making choices. If you didn't try to over ride the automatic [negative side] of the amygdala, you're not making choices, and that's a depressing thought."

NS: "What I see is that we're not always getting these overwhelming signals from the amygdala, as if we're about to be run over by a truck. But that we are getting much more subtle signals throughout the day at a sub-threshold level that can hijack our higher brain and decision making. to have an impact on us."

JC: "That's consistent with what I was saying. All of our senses get a reaction, and that's passed through our amygdala, and so there's data there that provides tint of influence. But that tint is primitive. When I'm most proud to be a human, is when I engage my frontal lobes to make a choice."

NS: "Do you think as a society, do you see evidence that we are controlled by those reactive processes in our brain far too often?"

JC: "Absolutely. That probably comes from a scarcity mentality that historically served its purpose when we were hunter-gatherers looking for berries and there might have not been enough to go around.

But one of the things that has allowed humans to survive, now seven billion of us on the planet- far beyond what people had predicted with dire consequences a hundred years ago- is this notion of cooperation which helps us to distribute resources a little better. That would be something different from what our amygdala alone clicking backwards might tell us about what is scarce and what threatens us.

So something has happened, and I think we're making progress."

AMYGDALA TICKLE #22- "Give"
Do something for someone else, without a thought of reward for yourself. Make a donation, help an old lady get on the bus, give a bum a dollar, volunteer for a cause.

FLTV

So, we've had black and white antenna TV, then cable TV, and finally digital TV. What's next?

TV in your own head- where you tune into the rest of the universe via the infinity reception of your own brain. Welcome to

Frontal Lobes TV.

FLTV makes an end-run around the endless debate about *this* versus *that*, around who the good guy is and who the bad guy is. FLTV just sees what IS from that big ol' camera in the sky.

FLTV is a leap of awareness-consciousness that suddenly transcends the problem making dualistic view of the world that is produced by the reactive and reptile brain, of "Me Versus You" that sees everything from the point of view of the Little Me Eye.

This truly un-evolved, repressive and fractured un-working brain process is replaced, *finally*, by

An evolved, connected, Whole Brain System

This transcended FLTV view of the world blooms a problem solving unified view of the UNI-verse that is seen by the Reflective Frontal Lobes Brain of "We Are All In This Together" that perceives everything from the You-Niversal Big Magic I that is all inclusive, like a **World Wide Interbrain**, where every brain talks to every other brain in quantum consciousness that moves faster than light, and doesn't require a monthly fee.

It is the WHOLE frontal lobes together, right and left hemispheres balanced in harmony, and *all parts* of the frontal lobes- not just the little bits your reptile brain might prefer to hijack in its own egotistical plans.

This is what the truly sustainable societies know, it is what the aborigines do, what the rainforest people do, what the un-electrified native peoples used to do before fast food big corporate daddy tore out the coconut tree to put in a convenience store, what the animals do, what the trees do.

It is finally what Truly Modern Man will understand.

At some point you give up your flat virtual screen tunnel vision, you put down your cell phone, you put down your Internet, you put down your Big Brother TV set- and you communicate and look all of your planetary neighbors right in the face, where no one can intrude and brainwash you with this one-way fact and that little twisted fact, all of which is no more than two-cents on the dollar.

At that point, people finally start to tune into FLTV and knowingly nod to each other:

"Hey man! Have you SEEN what I seen in my brain?!?" And everyone everywhere grasps and achieves telepathic consensus action from collective intelligence.

Then, Homo sapiens finally evolves into Homo novus (New Man)

Dr. Lawrence Blair is convinced that we are on the verge of a huge historic paradigm shift, in which the awareness of Brain Radar consciousness becomes the norm rather than the exception.

The old way of thinking is where humans see themselves and most everything else as individual parts, only remotely connected. The colossal shift in awareness on which humanity stands, as if we have one foot on a bridge that crosses into another universe, allows one to see the bigger picture, how any individual is simultaneously and perhaps more importantly just a part of another larger and bigger unified organism.

Dr. Lawrence Blair
Psycho-Anthropology, www.indonesianodyssey.co.uk, (Bali, Indonesia)

Dr. Blair's astonishing travels through Indonesia have been well documented in his many films such as *Ring of Fire*, and *Myths, Magic, and Monsters*. He has made his home in Bali for the past several decades.

LB: "I deal with consciousness in nature, how humans tend not to see what we think we're looking at, and how we absorb input by adjusting that input to match our expectations, our beliefs, and what we've already got what's in our heads.

My present interests at the moment are in regard to the psychology of paradigm shifts. There's no doubt about it in my mind: We're already moving into one that makes the previous ones look like Mickey Mouse paddling his toy boat."

NS: (laughs) "Please define this change for me as you see it…"

LB: "A paradigm shift is a fundamental change in the ground of meanings and beliefs that are collectively shared by the world. Each time a paradigm shift occurs it is a major stress on our psychology because we have to abandon grounds of previous meanings and beliefs.

This upcoming shift has to do with consciousness. There is strong argument that people are beginning to believe that consciousness is not at all confined to human beings, but is seen throughout nature and matter in a way that we have previously not suspected. To make a long story short, this new paradigm shift will have to do with how consciousness actually shapes the physical world.

There is quite a lot of data to support this from various fields, not only in neurophysiology and the psychology of perception, but in the earth and biological sciences."

NS: "I certainly see great evidence for this myself."

LB: "The new paradigm shift seems to be shaping up every bit as psychologically difficult for us to digest as were previous paradigm shifts were to the psychology of our ancestors.

We look back on things like the heliocentric system being introduced with Copernicus and Galileo and the stir that caused. And later ones in the 1800's, when we were informed supposedly by our common sense and religious beliefs that man was the center of the universe created in a fixed form like the other animals- until people like Darwin and Wallace pointed out that this wasn't the case. It was really shocking to people back then.

The other interesting point is that a shift seems to take about a hundred and fifty years to shed itself of its previous paradigm and to be collectively accepted as a new paradigm. Arguably, the evolutionary paradigm shift still hasn't fully been taken on, and as we know there are still many people who don't believe in evolution.

But a far greater one than that is occurring in the area of consciousness- the idea of a consciousness that is permeating all of nature, physically, but not only organically but also at the sub-atomic level."

NS: "Can you give me an example?"

LB: "There are many fascinating ones. For example, 'emergent behavior' where you have great many groups or flocks of birds or fish that begin to behave like a single organism. You've got things like the extraordinary Asian synchronous gregarious fireflies. This is a colony of fireflies that ignite and occlude at precisely the same instant with their own individual colony rhythm. They can light up a whole mangrove swamp like a flashing Christmas tree. They found out that its not just one of these things copying its neighbor by a hundredth of a second- it's absolutely instantaneous and simultaneous.

This has led some biologist to say that this is some sort of broader organism just reflected by individual discrete parts."

NS: "I think many people have given that idea a little consideration in the way that bees show a collective intelligence in working together in a hive, in a more recognized and conventional example."

LB: "Or *The Soul of The White Ant* by Eugène N. Marais who mentioned it years ago [1937]. But what I am talking about is a slightly different quantum level of that.

There's another wonderful creature that we filmed and put in *Myths, Magic, and Monsters* that is something like one of those toy Slinkys. It lives in the water. You remember those springs that you had as a kid that would walk down stairs?"

NS: "Oh yes- I've actually got one right in the other room with me- I still haven't grown up..."

LB: "It's about the same size as that except that it has a skin on it and it's moving along and it has a front and a back end, and an underneath part, and a forward digestive mouth and a rear excretive part. But you can beat it up with your hands in the water and the whole thing just disintegrates into a cloud of dust.

But as you watch, that cloud of dust reintegrates itself back into a single organism again, sometimes including some of the debris or matter on the bottom of the sea with it, and it begins carrying on and walking and behaving like a single organism all over again. And it happens within moments.

But it's a colony- If you look at it under a microscope it's a great mystery. Each little part of it has a positive and a negative end, a back and a front, but how do they know where to conglomerate and become specialized organs in this overall organism?"

NS: "That's quite incredible... What is the name of this creature?"

LB: "It doesn't have a name yet. They're very rare and have only just been discovered off the coast of New Guinea.

Here's something even more interesting- you have these massive stands of a single species of tree in the jungles of Borneo which under certain circumstances begin behaving like a single organism particularly in response to fire.

When fire approaches trees on the outer periphery of this stand, the trees will begin to turn the under side of their leaves towards the fire and they then begin to release the equivalent of a vegetable pheromone which reaches inwards many, many kilometers into the stand. This has the result of warning all of the other trees, and they modify themselves and begin crunching down and protecting themselves against fire."

NS: "That's one for 'Ripley's Believe It Or Not'."

LB: "From my point of view- and this is a very childlike view of the new paradigm shift- that what we may have on the planet are vast living organisms- giants that we don't see, because we only see the trees and not the forest. We only see their discrete parts rather than the whole thing, which throws a different light on ecology also.

There are other things that are suggesting a paradigm shift particularly in the area of quantum physics where we think consciousness is operating at the sub-atomic level. In fact it may be that there is nothing *but* consciousness and that all the rest of this is just an illusion."

Chapter 10

TWO SIDES OF THE SEE-SAW

BALANCE OF THE HEMISPHERES

According to Dr. Robert Neumann, head neurosurgeon at The University of Colorado Medical Center Hospital, real problems occurs when a person has either too much activity or too little occurring in the brain.

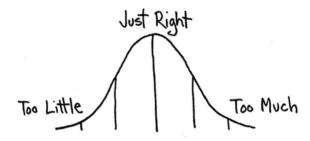

He describes how brain activity can be plotted on a graph- and the resultant constructive work being done by that brain can be seen as a bell curve, with the greatest and best work accomplished right in the middle, in between too little activity, and too much

We all probably have a good idea of what constitutes too little and too much brain activity, as we could intuit what would constitute too little or too much physical exertion. A coma would probably fit on the far side of too little activity right next to being in a semi-conscious drunken stupor. A panic attack facing a charging rhino or under the influence of a massive dose of crystal methamphetamine might constitute brain activity in the extreme. Neither end of the scale would be something a person would want to keep up for very long, nor would it be a good environment to be in for college entrance study.

But rather than try to state "what is too much" and "what is too little", I think it would be safe to observe that there are no hard and fast rules about what is right and what is wrong- because if you are crossing the railroad tracks and suddenly note a bullet train headed your way, a very excitable state of mind and very rapid movement of legs would be just the ticket you need.

On the other hand, BALANCE is an important concept to keep in mind when trying to self-regulate one's regular daily activities. And what one has to balance are two opposing ways of doing things.

Given that you have a right and left brain frontal lobe, each has its own way of interpreting and interacting with your You-niverse.

The balance that is just right between these two different compartments in your cranium will be determined by what your amygdala is telling you about your movement and direction towards survival- and how your Amygdala Compass is telling you how to proceed or retreat.

Just like you are aiming for that balance between too little and too much brain activity, you are also- whether you realize it or not- aiming for a balance, a homeostasis, between the two hemispheres of your brain.

Dr. Robert Neumann, M.D.
Head Neurosurgeon Intensive Care University of Colorado Medical Center Hospital

NS: "Please explain what you do at the hospital here."

RN: "I run the Neurology Intensive Care Unit specialized to take care of patients with central nervous system injury or disease. The central nervous system being the brain and the spinal cord. Patients can come in with anything from a hemorrhage to an ischemic stroke [lack of blood to the brain], to a gunshot wound.

The thing that I find most remarkable in the recovery of these patients is that doctors in the field of neurology and neurosurgery are taught classically that there are some very discrete areas of the brain that are responsible for certain functions- and usually that's true. For instance, I have the motor control for my right hand and my right leg are located in the cortex of my left brain. But there are some other more esoteric functions of the human brain, notably language and the expression of language and the perception of language that are classically seen to be located in certain areas: The expression of language in the left frontal area, and the perception of language a little further back in the left parietal frontal area..."

NS: "Wernicke's and Broca's areas..."

RN: "That's right. And often we see damage in those areas, in whom they have little or no speech deficit, or the speech deficit that they have isn't classic in terms of description, and in whom recovery we're finding is actually quite excellent. And I wasn't taught that in medical school. I was taught that these patients would have a permanent and unrecoverable deficit."

NS: "So, what is that telling you about the brain?

RN: "I think that there is probably a bit more in the way in which we term 'plasticity' and 'heal-ability', if you will, that prior to the ability to save these patients in great numbers as we're able to do now, we didn't fully appreciate."

NS: "One of the things that's drawn my attention over the years is the procedure known as hemispherectomy, and the manner in which people recover from that. In that procedure, fully, one-half of the cerebrum is removed..."

RN: "...and in some cases patients recover quite well, or better than one would anticipate."

NS: "But you don't lose half of your brain function."

RN: "I think when that sort of procedure is done in a highly controlled fashion that it is performed in, that patients do quite well. If there is damage holo-hemispheric- in other words a lot of damage to the hemisphere under non-controlled conditions, say in traumatic conditions- patients do much less well."

NS: "Could that be because you have damage to the underlying systems besides the cortex?"

RN: "Exactly. I think you have more damage under non-controlled conditions, and more secondary damage under non-controlled conditions that isn't fully appreciated under the current imaging technologies that are usually used.

So, as marvelous as they are, the current imaging techniques don't show us everything. They show us a lot, but they don't show us everything. And we are reliant on them. But often times the patient's image of their brain won't match how good, or in some other cases how poorly they look. I've found this to be persistently curious.

So in talking to families, the more that I do this work, the more that I know that *I don't know*. I just don't know."

NS: "This brings up an interesting question- We're all familiar with the saying that 'You only use 10% of your brain.'"

RN: "Yes..."

NS: "Which I recognize is not literally true, but I've come to recognize that it reflects some truth. Do you think that it is accurate in portraying that global feeling that we have this unlimited potential?"

RN: "That's a great question. The concept that we each have far-reaching untapped potential is a very tempting concept, because the next question it then leads one to is then, 'How do I get to the other 90%?'"

NS: "My question has to do with the relationship between this higher functioning in the brain in the pre-frontal cortex and the degree to which this fight or flight aversion is happening- and maybe more simply put, when we're clicked into our reptile brain as it were, doesn't that turn down our ability to learn? Conversely, when fight or flight is not activated, and we're feeling cooperative and loving, doesn't that increase our ability to learn?"

RN: "I do know that in terms of excitability there is a normal bell shaped curve for most functions and most learning capacity. On the left side of the curve when a human being is, say, very relaxed- hyper-relaxed- they don't learn as well. On the other side of the bell-shaped curve, when they're very, very excited, maybe we'll say stressed- they're hyperadreneurgic, common referred to as having a lot of adrenaline in their body, they also don't learn well."

NS: "You're talking about hijacking of the brain by the amygdala?"

RN: "Yes, they're in over-drive."

NS: "Now that could be both happy or frightened..."

RN: "Yes, that's correct. The emotional component isn't important. What is important is the fact that the subject in this case is hyperadreneurgic, they are flooded with adrenaline. They don't learn as well in either of those circumstances.

But there is a middle ground wherein there is just enough adreneurgic tone- excitement as you will- to enhance learning. But not so much that they're over-stimulated."

NS: "There used to be a 'pocket-neurology' way of looking at the left brain and the right brain. How much have we learned that such a view is an over-simplification?"

RN: "Oh, I think it has been over-simplified. I think in general those precepts, those Broadmann's Areas of function do still hold water, there's no question about it. But I think there's still more going on that that."

WHY TWO DIFFERENT BRAIN HEMISPHERES?

The first question one might ask is why do we have two different hemispheres that look and think differently to begin with?

Perhaps one can make an analogy with all of the other paired systems we use in our life.

Our body is bi-laterally symmetrical- meaning that we mostly have two of everything, on opposite sides of our body: Two arms, two legs, two eyes, two ears, two nostrils, and so on. The reason is that often, two is better than one:

With two eyes you get stereoscopic vision, allowing perception of depth of field.

Imagine the advantages of running and balance using two legs instead of trying to hop everywhere on a mono-pod pogo stick of one leg.

When was the last time you saw someone ride a unicycle across the winning line in the French Tour de France?

Two ears help you determine direction.

Have you ever played on a see-saw with yourself?

The meaning here is that when you have two things that work together or see the universe from a different perspective or fulcrum, you get a combined product that is better than one alone.

The brain, by being split in two, is in effect like having a team, a partnership, where one part of the brain can perform one function independent and different from the other part of the brain.

Another way to look at this might be to envision a task, such as hammering a nail into a piece of wood. Very hard with one hand, far easier with two. By having two hands, one hand can hold the nail steady or the block of wood, and the other hand can hold the hammer and strike the nail into place.

Also, don't neglect that since you have two hands, each side of your brain can instruct and manipulate each hand separately. This is an ability a piano player or a guitar player is intimately familiar with, and grateful to have at his disposal.

RIGHT BRAIN and LEFT BRAIN FUNCTION

Decades ago it was fashionable to see the opposing sides of the brain as having strictly different function. Now we know better, and understand that both sides of the brain contribute to many of the skills we formerly thought were limited to one side.

We have long understood that the left side of our body is controlled by the right side of our brain and the right side of our body is controlled by the left side of our brain. The exact way in which the functions of each hemisphere differ other than in controlling the opposite body side has taken a bit longer to understand.

The most recent way to look at the difference between the hemispheres shows some rudimentary differences that can easily be summed up in just one sentence:

The left brain sees a focused detailed small part of your You-niverse, whereas the right brain sees a fuzzy yet global view.

In this way, the left brain sees a localized, linear, separated pieces of an organized picture, while the right brain sees a non-local, non-linear, unified whole picture as a single unit.

In this way, the left brain is nearsighted and the right brain is far sighted. Although not strict or exclusive, we can make some generalizations about the differences between left and right brain hemisphere function:

LEFT BRAIN	RIGHT BRAIN
Effort	Effortless
Activity	Observation
Focused	Expansive
Universe as Separate Pieces	Universe as Single Unit
Talk	Silence
Doing	Being
Division	Unity
Time As A Progressive Line	Time as Now
Conscious Understanding	Unconscious Understanding

This right/left brain lateralization is not limited to human beings, but exists throughout the animal kingdom, because it allows an animal to perform one kind of mental task at the same time as another.

For example, a bird's left brain may be focused and looking for and pecking at worms or seeds at one small area on the ground, while simultaneously remaining alert of the overall surroundings, remaining vigilant for predators and other intruders in the distance from a broad area around.

Imagine how long you would last as a sparrow in a crocodile pond if you only had one side of your bird brain to use at a time.

Another example might be seen in typical tool usage: In our right hand (controlled by the left hemisphere) we hold a tool, such as a hammer, chisel, or screwdriver. This tool is manipulated to make changes to whatever we are working on. At the same time, with our left hand (controlled by our right hemisphere) we hold and observe the general state of the object we are manipulating.

DO BE DO BE DO WHOLE BRAIN POWER

Theoretical physicist Amit Goswami, observes, "The left brain is the 'doing' side of the brain, and the right brain is the 'being' side." He has expressed the opinion that whenever we limit ourselves to either "Do do do" or "Be be be", we get into trouble. His advice is that we change our tune instead to, "Do Be Do Be Do" to keep a balanced perspective and more productive state.

We can also make a reasonable and parallel observation about the bilateral symmetry of the rest of the human body. Each side of our body generally has a separate function apart from the opposite side. The combination of the two, in BALANCE, allows us to most effectively function in the world.

When the two sides are unbalanced, then we have more difficulty functioning.

If you have poor sight in one eye, the total vision suffers.

If you have a lame leg or arm, the total mobility suffers.

A good analogy could be seen in a car: In order to go straight forward, both sides of the car's wheels must spin with equal rotation. Both front wheels must point in the same direction. An imbalance will result in driving in circles-this is okay at the amusement park on a kiddie car ride, but few other places.

Interestingly, although the overall balance of any bi-lateral system is essential, alternate switching and opposition of both sides *also* counts.

For example, you take one step with one leg forward at a time, followed by the opposite leg moving forward by itself. This allows rapid progress, walking or running- certainly a better mode of movement than hopping, at least if you are not a kangaroo.

A bicycle is pedaled not by both legs moving downward at the same time, but by alternate strokes of each leg.

As it turns out, we alternatively also cycle and switch between using one side of our brain predominantly followed by the other. This happens similarly to the way in which we alternate through brain wave patterns during sleep, which occurs in roughly 90-120 minute cycles from REM (rapid eye movement) dreams to deeper unconscious theta and delta periods.

However, in the same way that you sometimes use your hands together, and sometimes apart, in the same way that you sometimes jump with both legs together, and sometimes run with your legs in opposite directions- the same holds true with your brain.

As it turns out, peak learning and comprehension happens when you have balanced activity happening in both brain hemispheres either simultaneously or alternatively.

And so, when we think about brain function, problems must arise when one side of the brain becomes overly dominant, when you see the universe and operate within your own You-niverse in an unbalanced way- when you predominantly use one side of your brain in great preference over the other.

Not only can you see unbalanced brain perspective and its profoundly negative effects in individuals, but you can see it in groups of people, and even in entire cultures.

Historical examples are easy to find: In societies and political movements that see people for their differences as separate pieces of humanity, rather than as human beings as having things in common.

The obvious deadly and horrific results of this one-sided (left brain) way of thinking is not difficult to reflect upon.

On the other foot, the dangers of naively seeing this one-sided view (right brain) with everyone as harmless and wonderful members of a global peace party of fairness and goodness is also apparent. Everyone, is indeed, not thinking of your best interest in their heart of hearts.

So what do you do?

You don't get stuck in a rut.

You look at the opposite way.

You use all of your brain, both sides, balanced in equilibrium, so you know when to run, when to talk, when to think of the future, the past, and when to think of the now.

You understand that sometimes you stand or jump with both legs together, and that sometimes you walk or run with both legs opposite- and you do the same with both hemispheres of your brain.

If one thing exists and is valid, then somewhere, someplace, the opposite thing exists and is also valid- The advanced modern brain explorer recognizes both sides of the coin.

The main thing is, humans have a brain that does an unlimited variety of things. You can use it for non-verbal, non-linear tasks and meditations, which are extremely valuable: Silent meditations, smile meditations, mantras, use of music, and other non-verbal things.

But sometimes you need it for just the opposite as well: Verbal, linear tasks, making lists, organizing things, re-ordering of things. You must certainly like reading and writing to some degree, or you wouldn't be reading this.

When you smile on *both* your right and left brain, you then are able to also use it in a balanced way that does not ignore those kinds of things that you also need, that might be just the opposite of the way one might prefer.

We have two legs because we can run faster using opposite sides of our body together. Imagine if you just used one side of your body- is using just one side of your brain any better?

It's best to avoid the "Dead Genius Syndrome", the inevitable catastrophe that follows when we only see the choices that come from one side of the brain.

Marie-Louise Oosthuysen
Brain Education Researcher, (Mexico City)

NS: "What is the most important thing that people can learn about the brain?"

MO: "The most important thing is that you have to be interested in what you want to learn. If you are not passionate about what I want to learn, it's harder for you to learn."

NS: "You're talking about positive emotion?"

MO: "Yes. The brain is wired for survival, first and foremost, most importantly. So, what I'm going to learn most quickly is what I need to know in order to survive- whether it's positive or negative. But it's so much easier if it's something positive and something that you're passionate about, and remember it because you're able to focus.

It means being able to concentrate and focus 100% on what you are doing at that moment. If you can do that, you'll remember it so much easier. The problem why we forget where we put our keys is because we are not paying attention. It's been proven over and over, that multi-tasking reduces our efficiency. You actually get less done doing more than one thing at a time.

You have to think about this too: The brain regularly goes through cycles. It is more or less ninety-minute cycles. People say that, 'I'm really right brained', or they say 'I'm left brained,' but it's not really true, because depending on which part of the cycle your in, that will determine which half or part of the brain is more active. Then in the next ninety-minute cycle, the other hemisphere will become more active.

The left hemisphere has a small, detailed oriented spotlight that lets you see a small, narrow area. The right hemisphere is more of a global spotlight, with a large area of perception. You naturally switch between those, in that cycle of approximate ninety-minutes.

Also, there's this: If you give a talk or teach a class, your audience can pay attention for ten minutes- and that's it. Then you have to switch gears and take a break, tell a story or a joke, or do an activity- then you can come back and talk for another ten minutes."

NS: "If there was one thing responsible in the human brain for causing all of these problems that we see around us- If there were one thing, what is that most basic root cause of everyone's problem that we just see as variations on that theme? What do you think that is?"

MO: "If there was one thing, I would have to say that we have to learn how to handle our emotions better. If I am more mature emotionally, the chances of an amygdala hijacking are next to nothing, or a lot less."

NS: "If all we do is react, then we're no better than a lizard or any other animal."

MO: "You have to be able to control your emotions so that you just don't over-react or react in a negative manner, and jump to conclusions.

You have to stimulate that neuro-pathway between the amygdala and the pre-frontal cortex. That's so that when you are in a stressful situation, you can stimulate that pathway so you can think things through very quickly and not over-react. That's why clicking or tickling the amygdala forward is so important- so you can practice stimulating that pathway.

Whenever you learn something, whether it's trying to improve your tennis swing or your golf swing, anything that you're trying to learn, whether its homework from school, learning to knit- whatever it is, you have to repeat and repeat and repeat. To remember that activity and use it well for the rest of my life, I have to repeat.

So- how am I going to train my brain to not over-react in an emergency situation? How can I prepare myself to think things through very quickly and completely, so I have all parts of my brain lit up and helping me with a good solution instead of just clicking back into an instinctual behavior and in the process killing myself and killing those around me? Freezing up completely, not knowing what to do? How can I prepare myself for that kind of emergency situation- to keep a cool head…"

NS: "That's what we pay firemen to do…"

MO: "Yes, and they keep practicing. They do drills on a daily basis, where they practice how to handle certain situations, with fire all around them- they prepare themselves for an emergency situation on a daily basis. So we have to prepare ourselves also for whatever may come our way.

We need to train our brain to look at all the other options instead of grabbing club or running away, perhaps off a cliff or into something else bad, very quickly. And the best way to do that is to click forward and to stimulate that pathway between the pre-frontal cortex and the amygdala. That is why it is so important to click your amygdala forward.

Myself, at first I would do it, ten, fifty, a hundred times a day- but then it became automatic, and I would find that in hairy situations I would automatically be calmer. If you make it a habit, when you do get into a sticky situation, you can control your emotions so much better.

And if you habitually learn to keep the pre-frontal cortex engaged by tickling your amygdala forward, you won't get into so many bad situations to begin with."

RIGHT BRAIN VERSUS LEFT BRAIN

Before advanced technology, people directly related to the natural world and were deeply connected to it. People understood how they were connected to the Earth and saw how they fit into it, because they could *not* divorce themselves from it and totally control it.

In the really old days, the rituals of magic and religion primarily modified your *perceptions* of how things were and how you fit into the bigger picture.

As humans learned to control the environment with technology using their left brain, they moved away from seeing themselves as an indivisible integral part of the world, as their right brain sees things.

Back when we were largely non-technological hunter-gatherer-farmers cave and hut people, it was "one person- one vote". You carried roughly the same weight as your next door neighbor.

As humans figured out how to make better spears, armor, and other technology that allowed small numbers of people to control vastly larger numbers of people and creatures, suddenly, the advantages of left brain mechanical power looked mighty tempting.

If you could figure a way to make a better weapon, suddenly, your survival and your happiness looked as promising as that big mountain gazing over the defenseless fruited plain.

It was no longer how big your heart was and how well you fit into your environment, but how well you outfitted your goon squad and how well you manipulated everything on the other side of the fence.

Left brain ruled, and things haven't changed much.

This has led to an imbalance which has reached its pinnacle at this point in history. We are at a tipping point where we are capable of utterly destroying the environmental balance and ourselves in the process, simply because we have operated under the left brain illusion that we can control everything.

The internal imbalance is reflected in the external imbalance. It is one and the same.

So how can these frontal lobes processes (left and right brain) of Cooperation, Imagination, Creativity, Intuition, and Logic be used in *evil* ways? Cooperation is indeed a frontal lobes process- But people actually cooperated with Hitler. How could that be good?

This is not because of increased frontal lobes processes, but rather, because of an *incompleteness* of it, an imbalance of the hemispheres.

As brain research has now made clear, if you don't connect with ALL of your frontal lobes, specifically, the medial pre-frontal cortex- you then act in anti-social ways, and potentially in pathologically harmful ways. It is the MPFC that allows you to empathize with others, and this is what keeps you from building and dropping The Bomb as an experiment on the family next door to see if it really works.

Those who don't include and think of others have failed to fully circuit into their *whole* frontal lobes. They are still stuck seeing things through their "Little Me Eye".

Those who view themselves as primarily separate from the whole have not transcended into the "Big Magic I", that includes *everything* and *everyone*.

"Cooperation" in the most limited sense doesn't mean to just go along with the survival of the local group. That will often result in various flavors of racism and fascism. That's the Little Me Eye view of things.

"Cooperation" in the truest and most powerful sense means to go along with survival of the really big group- everyone and everything. That's how the Big Magic I sees things.

"Cooperation" is in fact a directional vector: Your survival/happiness direction going in the same direction in The Big Picture, not just what is one inch in front of your nose.

One inch in front of your nose- that's what the left brain tends to see, that's a specialty of the left brain and the reactive reptile brain. However, a nearsighted brain is a brain waiting to be run over by the next truck that comes barreling down the street. To achieve long term happiness and survival, the left brain processes must be balanced with the right brain, to healthfully live in The Whole Picture.

It's not that the left brain can't see The Big Picture- it's just that it has a tendency to ignore it, before it has a chance to join up with its right side brother.

The corpus callosum only is problematic if the two halves cannot merge, like Oscar and Felix, the Odd Couple that always fights. When one transcends into Whole Brain Power, both hemispheres merge into a stereoscopic image.

Then, the left brain can explain and spell it all out with logic and words: Little Me Eye versus Big Magic I

It seems like that these days, every other brain researcher is telling you that "Happiness is in the left side of your brain" and "People who are happier show more left frontal lobes activity."

This is a very big trick played on the brains of very many predominantly Left Brained Folk who aren't seeing the Big Picture.

Dr. Richard Davidson, a big honcho in brain imaging and an expert in government grant acquisitions, will tell you that "a marker for happiness is more left frontal lobes activity". Unfortunately, Davidson soon contradicts whatever point he is trying to score, because he next informs us that the better you use your brain, the LESS active it becomes. Hmmm, let me check my brain... be right back... oh wait, I would be better if I use less. Gosh, now what do I do?

Davidson's observations are particularly irritating, because his sterling examples of expert happiness achievers are those guys whom he cites as having practiced for 10,000 hours in the Lotus position or more.

So, um, what did every single one of these happiness experts do FIRST to become Nirvana super stars? The same thing anyone does to get good at what they do: They spent the first 10,000 hours practicing and using their brains a LOT MORE so that then after 10,000 hours they could use it LESS.

Davidson forgets to mention that you can't get to point B (less) without having gone through point A (more).

According to Davidson's same logic, a marker for people who are FULL is that they don't EAT.

What about neuroanatomist Jill Bolte Taylor? (See http://www.youtube.com/watch?v=UyyjU8fzEYU) She had a stroke in the left hemisphere of her brain, and her left hemisphere virtually stopped functioning altogether.

According to Davidson and others, we should expect that she would be profoundly and instantaneously depressed and unhappy because she lost that left brain happiness machine generator.

Except that's not at all what happened.

What she immediately experienced was profound bliss and oneness with the universe. (Fortunately, what remained of her left brain managed to convince her to dial 911 for the ambulance.)

So, reconcile *that* one.

What neuroanatomist Taylor has clearly demonstrated is that right brain IS ALREADY THERE. It doesn't have to work nearly as hard as the left brain to experience happiness and unity with what is.

When you've already hiked to the top of the hill, you don't have to work so hard getting there. You ARE there.

That might explain why the monks and all the other "happy experts" have more activity in their left frontal lobes as compared to their right frontal lobes. In order to achieve a balanced state of happiness in both hemispheres, the left brain is playing catch up to the right brain

Maybe the problem isn't that we are not using our left brain enough, but rather that we have been using it TOO MUCH in an unbalanced manner all along.

Didja' ever hear about JUDO? That's when you use your opponent's energy to defeat him. Two guys in a ring, one guy knows judo, the other guy weighs 300 pounds and comes flying at the judo guy like a freight train out of control. The judo master just sticks out his pinkie and in one graceful little flip of his finger, sends the hulk flying over the ropes. He didn't use a fraction of the energy or force to win the match.

Maybe our left brains should just take a closer look at that judo match, instead of spinning its wheels for another 10,000 hours trying to become an expert at happiness, and learn from its right hand brother lobe.

Suzanna Del Vecchio
Oriental Dance Artist, International Belly Dance Instructor/Performer

Suzanna Del Vecchio is a dancer of international recognition, awarded 'Choreographer of the Year' in 1998, and 'Dancer of the Year' in 2001 by the International Academy of Middle Eastern Dance in Los Angeles.

NS: "How did you decide to become a dancer?"

SDV: "I had decided that I would be an elementary or secondary school teacher, but I wasn't happy with my college classes. So after three and a half years I dropped out, and moved to the northwest U.S. with my husband.

I had read a book in that late 70's about body work for women that had to do with the feminine revolution. It had stories about various women and their work, and I read about one woman who was a belly dancer, and I really liked what she had to say about it.

So I decided to take some lessons. I had some experience acting in high school and college, so performing wasn't new to me. But when I saw this dancer in a Greek restaurant to live music, I just knew, *that* was what I wanted to do."

NS: "Did you have that 'Eureka' moment when you realized 'This is what I want to do!'"

SDV: "When I read that book, that was it. I knew that I could do that, and that I *would* do it."

NS: "It wasn't something that you thought about in a rational, logical way- but it was more of a feeling that you had…"

SDV: "Exactly. I knew, 'This was *it.*' And a lot of dancers whom I know that became very well known professionals had that same experience."

NS: "That's 'The Frontal Lobes Pop!', when you get the information, and you don't know how you get it, but it's this 'Eureka Moment' and it changes your life in a big way."

SDV: "Yeah. When I read that interview, and then saw that dancer, I knew. But, you know, I am sure that at some point my left brain came in and said, 'Oh, you'll probably do this for five years, and then you'll have to get a real job at some point…' (laughs) But that's not what happened.

I knew I was also a good teacher, it just didn't turn out that I was a school teacher, but it was a belly dance teacher."

NS: "What have you observed about your students that separates the really good ones from the so-so dancers?"

SDV: "I think that people are either born with some grace, and then there are people who can learn the craft, and that's where most people fall.

There are a few people with a particular ease in the movement, and they are the one's who excel at it.

Now, even if you really excel, you don't necessarily get to the top of the profession. Because what I've found is that people who get to the top, really want to get to the top- and it doesn't even mean that they're talented."

NS: "You mean 'getting to the top' in terms of succeeding at the business and getting the professional recognition?"

SDV: "That's right. The best dancers are not necessarily commercially successful. There are many who *are*, and there are many wonderful dancers in the community- but for me, what I'm watching, and I've seen so many dancers, what I'm looking for is someone who touches me.

I'm moved by a dancer that has a soulful quality. Someone perhaps that has learned their craft so well that they can let go and take me to where I'm not even thinking about what they're doing technically, but that they're taking me somewhere else beyond that.

I've found that to be interesting a lot- teacher and performers who don't necessarily have that beauty or magical quality. Because to me, someone can have that magical quality and you can enjoy what they're doing, and they may not be the best technically. And then, you have the opposite, those who are really good technically, but don't do anything for you.

It's the total picture."

NS: "You've had to make a lot of decisions, in your life and your career? If there was a switch that guides you, what would that be? What's your compass?"

SDV: "I think about that a lot. It's not through 'thinking'. It's not through the left brain."

NS: "But you don't discard that..."

SDV: "No. I believe that there has to be a balance. But if I get nervous about something- I find it helpful to do *something* towards that direction. I try something different. I just relax and let go, and think about what it is that I want in my life.

I want there to be an ease and a flow in whatever it is. I'm asking myself, 'What is it that I really want to do?'"

NS: "Not focusing on the outcome, but on the process."

SDV: "It's like when I do choreography. I just move with it. I have a goal in mind, but I stay open to what it is going to be- I just stay open to that flow. It's about being receptive."

NS: "What's it like when it doesn't work? When things are screwed up?"

SDV: "It's grasping onto thinking that things she be done a certain way, not being flexible enough."

NS: "So how do you get unstuck?"

SDV: "I put myself in a position that's not the norm, or the opposite. Movement is the thing for me. Even if it's just getting out and walking, it helps me to relax. I've read enough that I know that if I stop grasping and relax, all will be well.

I'm an optimist. I don't let myself hang out in the negative for very long. I know the mind can go in that direction, but I won't let it go there for very long. I'll shift it around that negativity.

If I have anxiety, and know it's there, I'll get very still. And if it doesn't disappear than I'll do some movement."

NS: "This metaphor occurs to me: Your car gets stuck in the snow, and...

SDV: "You spin your wheels."

NS: "Well, you know me long enough that you understand this concept of tickling the amygdala. I think you've already explained it- but tell me yourself, how do you tickle your amygdala? How do you do it?"

SDV: "I move. That's it. To tickle my amygdala I move, I dance. Simple."

Ramon Kelley
Fine Artist

Ramon Kelley is an artist with an international reputation in numerous fine art painting and other media. He has earned the highest level of respect in his profession by critics and fellow artists alike. For forty years his art has been a staple in galleries, exhibitions, and in private collections worldwide.

He has repeatedly received the most prestigious awards from Oil Painters of America, the American Watercolor Society, the Pastel Society of America, National Academy of Western Art and The Allied Artists of America. He was elected to the Colorado Institute of Art Hall of Fame, the Pastel Hall of Fame and a charter member of the Museum of native American Cultures in Washington, D.C.

NS: "Do you have a definition of good art and bad art?"

RK: "Good art is what the artist does from his heart and his soul. The artist isn't afraid to gamble. You have to let what's inside of you come out naturally."

NS: "And not worry about pleasing someone else?"

RK: "Yes. In the old days, you could paint and be supported by the king or queen, or your patron. Today, you have to make your own money- in anything you do, so this can affect your work- and not necessarily in a positive way."

NS: "Tell me about your career."

RK: "When I was a little kid, I liked to draw like most kids. I just kept doing that. I remember in grade school, the math teachers used to yell at me because I used to draw in the books and on my papers, and I got in a lot of trouble. But I kept doing it because I loved it.

I think what made the big difference was that one day, when I was in the fourth or fifth grade, I did a drawing of a horses head, and I gave it to this girl. She said, 'Oh, this is beautiful Raymond! You're an artist!' And that just felt so good- I had a title: 'I'm an artist.'

In the service I used to do portraits for some of my buddies, for their girlfriends and such. So when I got out I decided I wanted to go to art school. There was nothing else I was dying to do. But I didn't have any money, so I just started painting on my own and started to collect books. I went through the books, studying, drawing. I painted a lot of bad paintings, but I learned how to paint."

NS: "So you're primarily a self-taught painter?"

RK: "Right. I never went to classes or teachers. But I could see what people were doing, I could pick up a magazine and look and learn.

Anyway, here was another thing- In my life, I also grew up in a pretty dysfunctional environment- an alcoholic father, the whole ball of wax- nothing we ever had was right. The little shack that we lived in, the windows were cracked, whatever-

Seriously. I really believe this- although we might be born a dwarf or have it hard, or something's wrong, we work harder than anybody else to prove that we're as good as anybody else. So we end up pretty good painters, singers, dancers, or whatever! Look at Toulouse-Lautrec! All artists are that way!" (laughs)

NS: "But there are people who have a tough time, and don't make that transition. Why is it that some people have adversity and they are able to overcome that and achieve some level of accomplishment and greatness, and other people never get past it? What makes the difference?"

RK: "There's no answer, there's not proof beyond just having the desire. You have a love, you have a desire. I had that desire, it was deep, it was heavy- it was like I was addicted. I was, and I still am. I love to paint.

But some people overdo it. Like this guy here [points to an example in an art book]. He got married and had a daughter, but he didn't have any time for his family. He was totally in love with his art. And that's all he did. He neglected his family. But for me, my balance is my family.

That's at least half my reason to paint- to support my family. The other half? To support my addiction to painting!" (laughs)

NS: (laughs) "So, what do you think is a crucial mistake that artists make?"

RK: "Money. Painting what the dealers want you to paint. If you're not painting what you want to paint, you're not creating any more. You would be surprised at the number of artists who do that. I can show you people in this magazine." [points to an art magazine]

NS: "What influences your work?"

RK: "Music, all the arts go together. I never really quite realized that until quite recently, in the past ten years."

NS: "Do you listen to music while you paint?"

RK: "A variety, different kinds of music. They asked Isaac Levitan, the great Russian painter- whom I study- and he said, 'If you're a painter, and you love painting, there's no way you can't love music. You have to love music.'

It's the same thing with the writer, and the singer, and the musician: If you do those things and you don't love painting you're just not going to succeed."

NS: "Why do you think that is?"

RK: "It's because all of the arts go hand in hand. It's because good music, or a good story, or a good movie- it just makes you feel so good. I couldn't exist if my total pallet was just painting. There has to be a variety of things to keep us going."

NS: "Do you paint if you're having a low mood?"

RK: "No I don't. That's a good point.

I think in all of the arts you have a period where you just burn yourself out. There's no more in there- you have nothing more to say or to paint. I learned that the hard way. The first time it happened to me, I fought it to no avail. Just burn-out.

And I remember talking to an artist friend of mine and he told me, 'You know, Ramon, the best thing you can do when that happens is just don't even bother to paint. Don't go in the studio, don't bother to look at art books, just watch some funny movies.' So, now I just ride the storm when it happens.

But most artists don't know that. A lot of them don't witness that, because during those lulls and depressions they just keep painting and their paintings are bad. What they're painting isn't worth a damn. They're wasting their time.

But, generally I feel good about what I'm doing. The more I paint- the more I learn. I feel like I'm just getting better and better and better."

NS: "Are you constantly seeking out new painters to study?"

RK: "All the time. You have to do that. Because if you don't, you'll fall in to that pattern of sameness. Even having my son here, who's also an artist, is helpful. I like that. I need the people element. I need people around me.

I like to have that slap on the back. But I also need to have that thing when somebody says to me, 'You know Ramon, that's not one of your better paintings.' I need that too. But I also like it when somebody pats me on the back and says, 'That's good.' We're human, we need support."

NS: "What' you're saying is a thread that I see among people who are good at what they do. They incorporate other people into their work at one level or another. They're not dependent on what other's say, but they are concerned- there's this consideration of another person's perspective."

RK: "To a certain extent, we use people to help us get ahead. I learn from my students. All you have to do is listen. That's the best teacher in the world: If you can listen, that's a total education for your whole life."

NS: "What do you observe in the art world these days, for good or for bad?"

RK: "One thing I notice lacking in traditional painting these days is that I don't see people inventing any more. People, for instance, will paint a landscape, and they just copy what ever else they seeing going on out there. People are doing things the way it's been done for a hundred years or more. Why not take that landscape and add a little abstraction? Why not add a little more color that's not out there in the reality, a little less color? Why not change the design, the composition? People don't do it, because they're safe doing what they do."

NS: "People play it safe in order to succeed commercially?"

RK: "Right. People are afraid to gamble. But in order to be a successful painter or writer, you just have to let yourself go. Just lay it down. Don't be afraid of it. And if you screw it up- start another one!" (laughs)

But if you just do what is safe and acceptable, you'll never succeed in what you're doing. You may make a living at it- but that's not what it's all about."

Sarah Jaeger
Functional Ceramic Pottery Artist

Sarah is an internationally recognized ceramic artist, specializing in handmade functional porcelain pottery for the home and kitchen as well as decorative pieces. She regularly teaches workshops at the university level, and has received numerous awards and fellowships for her outstanding achievements in her art.

NS: "What's your formal education and background?"

SJ: "I have a B.A. in English literature. For the first twenty-one years of my life I was very verbally focused in my education. I didn't think of myself as having any artistic ability at all. I had a really great liberal arts education.

Then when I was a senior in college at Harvard, I just got this notion that I wanted to take a pottery class.

I wound up at a local art center in Cambridge that had a pottery studio in the basement. I signed up for this class, and I have to say, it really grabbed me. I found myself cutting my Chaucer classes to go over there and mess around.

Really, for a number of years I tried other jobs that I thought that I had been educated for."

NS: "Where your parents interested in the arts when you were growing up?"

SJ: "No. But there was this: My parents bought this place way out in the country, this beautiful old farm house. And I have very powerful memories of that house, where I lived until I was eight. And my parents collected antiques in New England.

So we had this hand-made house that was full of hand-made furniture, and we used everything. I still have some chairs from then that were from the 1700's. These objects were really treasured, but they weren't precious. We used to put our feet up on the table, and stuff like that. The things were used lovingly and carefully, but also casually.

There was that whole sense of useful things that were made with somebody else's hands- handmade functional things that were a part of your life. I think that was a really big part of what drew me to doing what I do."

NS: "Can you elaborate?"

SJ: "Well, say it's a cup that has a job like any other cup. But the fact that it's a cup that I put myself into while I'm making it, it communicates that at some level when somebody is using it. I mean, we can sit in a chair and think, 'Somebody labored with their hands and not very many tools to make this. It's a really beautiful thing that works really well.'"

NS: "You're recognizing the value of artisan goods as opposed to mass produced goods... At a certain level we all use mass-produced things..."

SJ: "I have mass-produced things that I like a lot."

NS: "But when you're talking about hand-made pottery or furniture, there's a secondary level of appreciation that you can have with that kind of object, because it has a uniqueness or personality to it that you wouldn't find in a factory produced object. And that there's an intrinsic level of enjoyment you can have with such a thing because of how it's made."

SJ: "I definitely feel that way."

NS: "Where do you get your inspiration from? What has been instrumental in shaping your work?"

SJ: "Even when I was back at Harvard, I had started looking at historic pots. I had been to some museums in Boston. I was first interested in Japanese Tea Ceremony ceramics. In Japan, ceramics are as honored as any other art form. So I did a lot of reading in that area, which helped me to realize that these objects that people might make out of clay carry as much meaning in the culture as any other object."

NS: "What about your travels- has that affected your work?"

SJ: "Definitely. Some of the places I've travel has been about observing textiles, folk art, and indigenous art, a lot of things made by pre-industrial societies. I have a strong feeling of kinship with a lot of objects made in cultures that are not industrialized. These kinds of cultures are where utilitarian every day objects are still made by hand, so there's not that disjunction between beauty and function.

We've kind of gotten to that point after industrialization where people say, 'Oh, that's just too pretty to use,' or if somebody has spent a lot of time embroidering a shirt, they're not going to wear that shirt when they work on the farm- they'd wear it to a fancy party.

But in some cultures, someone would spend a lot of time embroidering the shirt, and then they wear it on the farm."

NS: "So how does that relate to what you do?"

SJ: "Well, a machine could make a cup way faster than I can. But, it matters to me a lot that I can spend this much time making something that people can use. That's where I feel a relationship to those cultures, and where our culture has gotten away from that. The hand made objects in our society have become luxury goods, not every day goods. I think that's too bad."

NS: "So your work is in that middle ground, where you create functional pottery that has an esthetic beauty and thought behind it, but if you use it and drop it, it's not the end of the world."

SJ: "Exactly."

NS: "In terms of creativity and new ideas, you must certainly like most artists come up against a point where you say, "Now, what's next? I'm burned out doing this same thing..." What do you do at that point? What do you do when you need to refresh yourself?

SJ: "My formula, if anything, is just to make myself go to the studio and keep at it- and then stuff happens. Sometimes it happens sooner, and sometimes it doesn't happen until later. If I'm not in the studio, it definitely won't happen."

NS: "When I talked to Ramon Kelley the oil painter artist, he told me that when he didn't feel inspired he just put his brushes down and did something else until he felt motivated again."

SJ: "I guess *I'm just the opposite*. Where I get my ideas is in the studio- so I have to be in the studio doing something. I have to have clay in my hands to get the neurons firing in my brain where the ideas come from.

It may be the nature of what I do. The actual material I work with carries much more weight than the concept of what I do."

AND WHAT ABOUT THE REST OF US?

Much of modern life, especially urban living, is dominated by Left Brain activity. We work in cubicles, sit at keyboards for hours every day, file paper after paper after paper.

This predilection probably has given rise to the notion that the Left Brain reigns king, even by some brain scientists who should know better.

But more likely, it is that we have become, to our own detriment, increasingly dependent upon this hemisphere of our brain, to the wanton nearsighted neglect of holistic perspective and feeling.

And this results in Brain Burn Out, that most of us can immediately relate to on a daily basis.

The cure?...

AMYGDALA TICKLE #24- "Right Relief"

When you've been taxing your Left Brain too long, before smoke begins to pour out of your ears- take a Right Brain Relief Break.

Balance your brain, and get twice the mileage at half the price.

Walk the dog. Listen to music. Start dancing. Ride your bike. Paint a picture. Play your bass guitar. Cook some string beans.

Ahhhh.

WHO'S RIGHT? AND THEN, WHO'S LEFT?

The idea that happiness and fulfillment is indicated by more activity in the left frontal lobes is nearsighted microscopic crumb on the pie plate of life.

As clearly illustrated by the preceding artists- a dancer, a painter, and a potter- whose entire happy lives revolve around right brain activity, the idea that a preponderance of activity in just one half or the other of the human brain signals superior happiness is a ludicrous notion.

Perhaps Monty Python member Terry Jones has got a much better handle on what indicates a better working brain than your neighborhood brain scanning machine...

Terry Jones
Founding Member Monty Python's Flying Circus, Film Director, Writer, Actor (London)

Terry Jones is famous throughout the world for his groundbreaking role as a founding member of the illustrious Monty Python comedy team, as a director of many films and television programs including *Monty Python and The Holy Grail* and *The Life of Brain*, as a comedic actor, and as a writer in many genres including adult and children's books.

TJ: "Hi Neil."

NS: "Hi Terry. First of all I want to say that I've been the greatest admire of yours since I was a teenager watching Python in my mother's kitchen many long years ago."

TJ: (laughs)

NS: "...I can't tell you how pleased and honored I am that you've agreed to take some of your valuable time to speak with me on the subject of the human brain and human potential as you see it.

So anyway, I want to start out with a question that I'm sure my readers would like to know your feelings on- and that is, I want to know if you prefer a powder or a liquid to get your socks clean and to get your whites their whitest white?"

TJ: "A powder or a liquid. Hmmmm. Well, I think neither really, because you can't write if you're under the influence of anything, I don't think."

NS: "Alright. Well, that answers that question.

You've recently written a book called *Evil Machines*. I'm wondering if you might elaborate your feelings about how technology has affected human thinking..."

TJ: (laughs)

NS: "...Either positively or negatively..."

TJ: "I suppose the computer is the biggest influence. When I used to write long hand rather than type, I used to write long hand on the right side of an exercise book. And then if I was changing anything, I would cross it out and write it on the left hand page of the book. That was pretty efficient.

Then, when you came to type out the long hand, you would start to reduce out the words because you didn't want to type for so long. So you shortened sentences, and that was an editing process in itself. And then reading stories to my children was also an editing process- you could tell if something was working or not.

But then with the computer I tend to just write things and then read it through and think it's okay and I think it's more verbose, there's more verbosity in it when it's on the computer. There's not that final editing process when you're typing it out."

NS: "So do you think the computer has been more of a hindrance or a help?"

TJ: "Well, I don't know... It's wonderful in its way, for like writing screenplays. It's brilliant with Final Draft. But, I don't know. It allows the free flow, but it's a problem with the editing process. I haven't really worked out how to deal with that."

NS: "Ah ha. Well, I'm writing quite a bit about creativity and that process inside people's heads. I'm wondering if there's some sort of switch that perhaps you tickle in your brain to get a good idea, or to bring about that creativity. If there was a switch, how does that work for you?"

TJ: (laughs some more) "I wish I knew- I wish I had the answer!"

NS: "Do you think that creativity is more of a bubbling up of the unconscious?"

TJ: "Yeah, I think so. For me that's what it is. I don't plan anything out in advance. For me the excitement is going to the page of writing the next morning and discovering what's going to happen next. That, for me, is the great drive for writing. Some times it works easy, and some times it doesn't.

I've just written an opera for the Royal Opera House- the libretto for the Studio Theater there- and that just wrote very easily. I directed it, and we put it on in April last year. It was just a joy to do. Anne Dudley was the composer. She's a film and TV composer and very well regarded. Stuart Copeland, the drummer for The Police did a piece as well that was part of the evening. It was all great fun.

Our piece was about a hour long, and it was about a wonderful doctor who's patients all love him, and he has a great cure record. But the medical council tells him that he has to stop practicing... because he's a dog."

NS: (laughs)

TJ: "But the patients say, 'Well it doesn't matter that he's a dog, he's just a wonderful doctor!'"

NS: "What's the name of this?"

TJ: "It's called 'The Doctor's Tale'"

NS: "Do you spell that T-A-L-E, or T-A-I-L?"

TJ: "Tale. It was suggested the other way, but I didn't like the pun."

NS: "You mentioned two words when you were describing your creativity- One was 'joy' and the other was 'fun'.

Do you think these are guideposts that you depend upon?"

TJ: "Yeah! I think I want to be happy when I'm creating something. It's a great joy to write.

I do know that Douglass Adams [*Hitchhikers Guide To The Galaxy*] hated writing. He just feared it, and was always behind on his deadlines, and he was a friend of mine. He had asked me to do a voice for a thing called *The Starship Titanic...*"

NS: "Yes, I'm familiar with that."

TJ: "Well, what happened was that Simon & Schuster had given him two million dollars (laughs) to produce something, and he hadn't done it seven years later. He still hadn't done it."

NS: (laughs)

TJ: "And so he said, 'Could I just do a video game of it instead?' And they said (grudgingly) 'Oh, all right...'

Then they decided they wanted to have a book as well, and the commissioned a science fiction writer from the '60s to do it. But Douglass hated it.

I had just read his film treatment for it, which was about twenty pages. And he asked me if I would be interested in writing the book for him. (laughs) And it was such great fun! It was one of the most enjoyable things! It was like when you see people writing in a film- The story was all there, the characters were all there... I sort of messed around with the story a little bit. But I would just do a chapter and go, 'Oo, I wonder what happens next?!' (laughs and laughs) We had five weeks to go and I did it in three weeks, just working in the morning. It was just... er... hoo ha!... lovely..."

NS: "I can relate to that, as a writer myself, because often I'll have to drag myself to the keyboard- but then once I get going, I get hooked and I forget about time and the effort, and I get into the flow of it and it becomes effortless."

TJ: "Yeah, yeah. If it's working, it's wonderful, and time never passes so quickly as it does when it's really working."

NS: "I'm writing about this phenomenon called 'Popping Your Frontal Lobes'."

TJ: "Yes..."

NS: "It's this 'Ah ha!' moment, this 'Eureka!' moment where you have this breakthrough, where you suddenly stumble upon a solution for something that's been gnawing at you."

TJ: "Hmmm..."

NS: "Have you ever had that kind of moment?"

TJ: "I think so. I'm thinking back on the children's books. I remember writing *Eric The Viking*. I was really stuck when they were down in the middle of the Earth or something. Then I suddenly had an idea about something that had me stuck for a couple of days.

With me, generally, I don't notice it happening. I get to the end of a morning's writing and go, 'Oh, Ha ha! That's quite good! (laughs) I think I've got that!' (laughs)"

NS: "Do you think your best work comes about effortlessly as opposed to forcing something to work?"

TJ: "Yes, I think so. For me it does, yeah."

NS: "In your work, both with Python and post-Python, obviously with Python there had to be quite a bit of cooperative thinking and behavior, and I know the history and I know there was a tug-o-war. But since then, how do you view cooperative behavior in regards to creativity?"

TJ: "Well I know one of the joys of novels is that what you write is what the reader reads. So it's like when you're reading in bed at night, it's like the author is whispering in your ear, that sort of thing. That's a very special relationship between the author and you.

When I was with Douglass Adams doing publicity for *The Starship Titanic* it was so wonderful to see how his readers reacted to him- they just knew him.

However, with film, it's a different matter. I enjoy making movies, but it's a cooperative venture as far as I'm concerned. I think Terry Gilliam would give you a different opinion. To him, he just has to get his vision on the screen. But for me, I have an easy rein and try to get everyone involved."

NS: "You've done a number of films on history. Recently I've just watched "The History of 1" which I thought was a fantastic film about math and numbers in civilization. So you have an interest in the history and culture of mankind in general. Do you have any thoughts about the high and low points in civilization and how it might relate to the way in which people fundamentally use their brain?"

TJ: "Ah... I don't know about high or low points in civilization. I really don't know very much about history. I know a little bit about the late fourteenth century. When I did *The Barbarians*, for example, it was because I hadn't studied the classical world and I didn't know anything about the Romans, and so there was a great learning curve in that. (laughs) And I didn't like what I found out!"

NS: "I just began watching that yesterday, '*The Hidden History of Rome*'..."

TJ: "Oh, that was a different thing. *The Barbarians*, I think, was a series for the BBC."

NS: "Oh. Okay."

TJ: "It was about 'barbarians', what the Romans called 'barbarians'. And it appeared that they were more civilized than what the Romans actually were themselves." (laughs)

NS: "This is a key point that I'm addressing, that has to do with civilization, a truly civilized culture and the relationship between it and technology and information. People make the assumption that more information and more technology makes for a more civilized culture. And it sounds like you're saying that this is not necessarily true."

TJ: "Yeah, well, for example the Celts, they didn't write anything down really, because they wanted to cultivate the art of memory. There were the Celtic scripts, but they were afraid that if they wrote everything down that they would lose the power of memorizing things."

NS: "These were who, again?"

TJ: "The Celts."

NS: "The Counts?"

TJ: "Celts."

NS: "Counts?"

TJ: "C-e-l-t-s."

NS: "Ohhhh. Ohhh. As in Celtic!"

TJ: "Yes. Celtic."

NS: 'Okay, I understand."

TJ: "Basically they were inhabiting France and England. But I don't think that human history proves that we are any more intelligent than... Aristotle was saying things as far as he could understand about the world in 360 B.C., and I don't think the human brain has gotten any bigger.

I think Plato and all of the Greek philosophers were just as bright as anybody could be."

NS: "It sounds like the point of your opera is in some fashion about that, the dog..."

TJ: "Well, if the dog is a very good doctor, then why shouldn't he practice?" (laughs)

NS: "I have one final question... You've had a great deal of success, critically, artistically, and I shall assume to some degree financially and materially, and probably socially as well. As you look back on your experiences in life, for you, what stands out as being most important and valuable among all of these measures of success, as society defines it?"

TJ: "I think that it's just *enjoying what you do*, really.

I feel so sorry for people who work nine to five and don't enjoy what they're doing, although they have to do it because they're wage slaves and they have to earn of course.

We were very fortunate in that Python has continued to provide an income because of the ABC [U.S. television network] court case in New York. The BBC ended up having to give us the rights to the shows. It was a very extraordinary situation where we owned the rights to the BBC shows, and that provided for us. And *The Holy Grail* came back to us. And we won the rights to *The Life of Brian*. It's only *The Meaning of Life* that we don't own the rights to.

So the movies and the TV show has provided an income over the years, so I could do what I liked. I've done a lot of academic work, these things about history. I've been thinking about Richard the Second, and Chaucer."

NS: "It sounds like what you're saying is, 'enjoying your work' is what you value most..."

TJ: "It's such a blessing to be able to enjoy your work, yes."

Chapter 11
HOW MUCH OF YOUR BRAIN DO YOU *REALLY* USE?

How often have I heard a brain "experts" say that, "It is a *myth* that you only use 10% of your brain" and "You use *all* of your brain all of the time."

How often? It now seems about as often as I see a new dandelion pop up in my back yard.

Years ago I finally learned not to take anyone's word for anything without first thinking about it. This would especially apply to "experts" of any kind.

To start with, any person, scientist or otherwise who claims that "you use all of your brain all of the time", is inferring that they can determine how much brain you are using.

How can *anyone* determine how much brain you are using? They can't, and they never will be able to know.

"The brain indicates its powers are endless", from "The Brain-Mind Problem", lecture delivered by Nobel Laureate Sir John Eccles at the University of Colorado on July 31, 1974.

Eccles spent his life wrestling with the subject of the human brain and mind and its potential, outlined in many books such as *How the Self Controls Its Brain*, (1994), or *Neurophysiological Basis of Mind*, (1953). It was Eccles observation and belief that the human brain has infinite potential, because for all practical purposes the human brain taps into consciousness and a system that exceeds what is inside any individual tissue brain box alone.

In other words, even a methodical and careful scientist like Eccles knew the human brain has an infinite capacity to know the universe, because the workings of the human brain is not limited to what might be stored or occur at any given moment in one's privately owned gray matter cranial shoebox.

A good analogy can be seen by comparing your brain to a computer hooked into the internet.

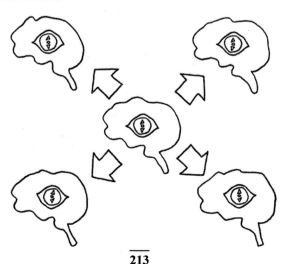

What your computer knows and can do with the infinite information accessible to it on the World Wide Web is incalculable. The working "consciousness" of your computer, is in fact, as big as its connections to every other computer connected to the Internet, consisting of millions of other computers, with mind boggling amounts of data- all of which is growing every second of every day without bounds.

Your computer is not limited to what software and data it may have on it at any given point, since you can always install new and more powerful software and data at any given time. Further, your own computer can hook in "the cloud" onto other computers and systems non-locally to directly benefit and utilize from that exterior storage and processing power.

Your own little dinky $300 computer can instantly connect online and play movies that were created with supercomputer technology made costing hundreds of millions of dollars. Further yet, you can watch an unlimited number of these brain numbing online videos whenever you choose.

Your computer, linked to all other computers on the Internet, is just one eyeball – but it's on a much bigger super-computer than itself far, far greater than what is sitting on your desktop. Your personal view screen is just a single nodule on a billion head beast that makes Medusa's wig look like a spaghetti string mop. And you can see what all the other heads are seeing- as they can of your view.

However, as impressive as the Internet is – it is nothing compared with your brain and its capacity to hook into not only what is online, but to every other human brain on the planet, and in addition, to every other item in the Whole Universe Super Duper Department Store.

Your brain is as big as the cosmos. Every flower you smell, every piece of cake you eat, every star you look at, every annoying pebble you get into your sneaker is a bit of information that your brain can juggle and re-juggle anew. There is no limit as to how far you can take it or how many bytes you can bite.

Einstein had the very same size brain and brain engine as everyone else, Mozart, Madam Curie, and Britney Spears included. Yet all of these people were able to do something completely different with their brain. They weren't limited to display the same dance moves as Britney Spears- because it's not in the hardware alone.

How much work can a brain with an infinite resource do?

An infinite amount of work. It can not be determined in any fashion or shape.

Despite this fact, all kinds of white lab coated brain researchers can be heard uttering that same utter mumbo jumbo nonsense any time of day or night:

"You use all of your brain all of the time."

Hogwash.

The brain is capable of doing many things- sometimes we sit and hum a tune, sometimes we calculate math, sometimes we have a memory, sometimes we watch clouds and think very little.

The degree and intensity to which neurons communicate with each other to create "thoughts"- silent or otherwise- varies from moment to moment. Sometimes we are alert and our brains are filled with ideas. Sometimes we are in deep sleep. Different parts of the brain are more active than others from moment to moment- this alone indicates that you do NOT use ALL of your brain all of the time.

There you have it! On every single fMRI (functional MRI) and PET (positive emission tomography) brain scan ever taken- proof positive that you do not use all of your brain all of the time.

Because if you did, such a machine would be completely useless in the first place. If you used all of your brain all of the time, every part of your brain would be uniformly lit up on the brain scanner continuously. The only reason for these brain function machines to exist in the first place is to detect and indicate how different parts of your brain are working at different intensity from other parts of your brain at different times.

And even that is only to the degree such a machine can approximate.

It is true that all your living neurons are always in *some* state of activity- they are alive and receiving oxygen and nourishment at *some level*.

Is this what is meant by "We are always using all of our brain" ?

That hardly says anything at all, and is about as insightful as saying "We use all of our muscles all of the time". Clearly, we do not *use* all of our muscles all of the time.

We certainly do not tap into the full potential of our leg muscles just because are muscles cells are alive all the time. Nor do we tap into the full potential of our muscles when we drive our minivan to 7-11 to fetch a greasy yet delicious bag of potato chips.

All brain scanners show is that *some relative* level of activity is occurring in *some relative* part of the brain for *some relative* amount of time.

Brain scanners can never tell anyone *how much potential* work the brain is capable of doing, nor can brain scanners indicate even remotely *the quality of the work.*

In a car, potential varies if the trunk is empty or full, whether the back seat is loaded or unloaded, whether you have air in your tires or you are riding on your rims. Similarly, there are a great many factors that determine exactly how much work your brain will perform at any given moment, regardless of how much glucose it is consuming or electrical current your synapses are firing off indicated by brain scanners and EEG machines.

If you drive your brain into another car at 70 miles per hour, is that using all of your brain? Many people test this theory every year.

Besides the inherent infinite capacity to know the universe in an infinite variety of ways, you cannot ignore the fact that you can sit in your chair all day and fill it with lies, false assumptions, and bad information.

You can almost certainly attest to that fact from thinking about some of the people you knew in high school.

Besides all of this, brain scanning researchers all make one distinctively bad assumption:

All brains work the same.

No two brains work in the same way, except in very general terms.

One recent notable laboratory study performed by Steve Quartz and Annette Asp at the California Institute of Technology illustrates this well. ("Social Valuation of Cultural Objects and Neural Patterns of Activation in Social and Self-Referential Structures")

The research was designed to determine how the brain distinguishes desirable from undesirable consumer goods in their "What is Cool?" experiment. As the brain scan results demonstrated, different people used different parts of their brain to distinguish their preferences. Only one-third of the subjects showed similar brain region activity in the frontal lobes to determine what they liked from what they disliked.

According to their own published results, two-thirds of the subjects did not show increased activation of their frontal lobes to make their decision.

So much for people using all of their brain all of the time.

It appears that the folk saying that "You only use 10% of your brain" is an intuitively correct metaphor to indicate that no matter how much brain you think you are using, you can never approach "brain completely used"-

At least not until your funeral day when your relatives can honestly say- "Yep. Brain all used up."

The idea that anyone- including yours truly- uses more than 10% of their brain is overly generous. Any percentage of infinity is by definition, an infinitely small piece of the pie.

The real joke is on those people who would erroneously claim that we all use all of our brain all of the time. What a horrific and pessimistic attitude that is. What hope do any of us have on this poor planet if everyone is already using all of their brain! Yow.

The truly optimistic view is to observe correctly that we only are using an infinitely small percentage of an infinitely limitless potential that resides inside each and every brain, a considerably reasonable and thoughtful perspective of the possibilities that reside inside our craniums.

Certainly, a more relevant and perhaps answerable question is-

"*How* are you using your brain?" And even more to the point-

"How are you Tickling Your Amygdala?"

Paul Epstein
Independent Record Store Owner "Twist and Shout", Businessman, Musicphile

Paul runs one of the most popular independent record stores in Colorado, and it is perhaps one of the most successful in the United States. This is quite a feat given the current state of the record business, where most of the major record retailers have dropped like flies at the end of summer. His business continues to thrive in a challenging economy against all odds.

NS: "How did you get into the record retail business?"

PE: "The opportunity just fell into my lap. I had been a high school English teacher, but I was known as a huge collector of music. That ended after a few years and I went back to teaching.

My wife recognized that I really didn't seem suited for teaching, and again, another opportunity just serendipitously came my way, this time in Denver, and we bought a record store that was up for grabs."

NS: "This strikes me as important, that you listened to someone else who seemed to clarify what you might have dismissed on your own. Is she different from you in some ways?"

PE: "Yes, very. She is very different from me, but complimentary. She listened and observed, and empowered me to want to make that change from teaching- and this second opportunity just presented itself to us, and we did it.

We were on a walk one night, walking down Pearl Street up to where I knew Underground Records was at, and I said to her, 'If I'm ever going to get another store, it's going to be like this one...' And as we get to the front window, there's a sign up that says, 'Seized: Tax Auction', and it's on the first day of our spring break at school. And we said, 'Let's go.'"

NS: "Were you a musician?"

PE: "I was a fanatical music collector. I played guitar and loved it, but it was clear that my path lay in archiving and studying, explaining.

It came from my father, who was an author, a professor, a collector, and music fanatic- all of the things that I turned out to be. (laughs) So I got the bug from him, and also my brother who was also a big collector.

The 1960's and mid '70s had a profound effect on me. I became obsessive about the music, in a very solitary way- reading, listening, amassing, understanding. I was not a very social person, so it was an inward thing for me."

NS: "It's interesting, because we tend to think of music lovers as right brained special oriented people. But you integrated this whole other left brain thing of cataloging and organizing at an intellectual level."

PE: "It was both of those things. I was initially excited by that first tier of it, the hedonistic pot smoking, acid taking- but on the other side I had this very intellectual teacherly approach."

NS: "Well, you were an English teacher."

PE: "Right. It was for me about trying to get my arms around it, controlling my environment. In my mind, that's what cataloging is all about. I've said to people, 'I'm the integrated man', what I do at home is very similar to what I do at work. I'm this obsessive, orderly, cataloger of music in an academic way, but I'm very close to the hedonistic- I like rock and roll.

But it's everything to me. It's very difficult to separate myself from the soundtrack that is constantly playing around me. But not now, because I didn't want to wreck your interview recording. (laughs) Normally I would have the record playing going the minute I walk in the house."

NS: "You've seen a lot of music careers that span a long period of time. What do you think about people who are successful at the beginning of their career, and then lose that spark as they go along?"

PE: "Well, I don't find fault. But I think that everybody has so much in them. This has been proven over and over and over. Everybody has so much in them, and normally at some level it's connected with youth and that initial drive. Some people have transcended that- although not many.

It's not a cultivated thing- either you have it, or you don't. It's the ability to tap into the inner child. It's the ability to hold onto the thing that initially excited you about your chosen field. I think that at some level, Neil Young, Bob Dylan have it- although there would be some people who would disagree with me. I think there are a few artists who manage to hold onto that initial thing that excited you.

For me, in my business- that's the key that's allowed me to hold onto my success. It's been two things: Not being afraid to change, but also, one, to do things in an adult manner, to complete things in a linear manner in a correct fashion. But two- ultimately more importantly- to hold on to the very passionate, child-like side, and not lose that- which every single thing about the other side encourages us to do.

Everything in the adult world encourages you to lose track of what originally excited you as a kid."

NS: "You said something key, which was, 'Not being afraid to change'. I think people say, "Oh, I was successful because I did *this*, so let's just keep doing that same thing.' And that's what trips you up."

PE: "In the business that I'm in, I remember early on I made this discovery that the musicians that excite me the most are improvisers: Charlie Parker, and Coltrane, Jerry Garcia- people who learn their instrument to the level where they could improvise, which I know as a musician you totally understand.

I had this revelation early on: That 'retail' is this job's improvisation. What I do, is *that*. I am like the musicians that I admire most. My job every day is different, and that's the way I want it. It gives me all this room to improvise. I'm the master of my own destiny. And as long as I'm not afraid to go down dark alleys and change and do things differently and master the situation...

You said something to me earlier, you said 'Once someone figures out what works, they don't want to change what they do.' People are tremendously lazy. *Tremendously.* I've had kids, children, grandchildren, students, and *hundreds* of employees.   And thousands of customers. So I've observed human nature and how people behave.

People are tremendously lazy in all walks of life. That goes hand in hand with not losing touch. Because if there's one thing that kids are not, they're not lazy. They're always going, 'Whoaoaoaoa!!! Ayaayayyyaaaa~!!'" That's the state I've always wanted to remain in.

That's the key there- it's been cast in adult society to not be lazy- but everything about adult life tries to shut you off from that, from the very natural and intuitive way of approaching problems.

But it's a very childlike thing, it's the most natural intuitive thing to be excited about what you are doing, to want to go in there and not be afraid to mount new challenges. That's where it's at."

AMYGDALA TICKLE #25- "Play With Your Own Tickle"
Play like a kid with your very own Amygdala Tickle.

What do you like? Angels? Laser Beams? Butterflies? Magic Wands? Quantum Energy Flux Rays? Tactile Fingertip Tickles? Heart Waves? Poetry? Guitars? Paint Brush?

You name it. Make up a way to pleasurably stimulate the goodie-spot inside your brain. Use sound, imagination, touch, smell, words, music, walk in the woods—you pick. It's your brain, and your amygdala. You know what works best for you.

Then try someone else's method- Perhaps combine, mix and match.

The sky is the limit, the options are infinite.

All you have to remember is CICIL

Cooperation-Imagination-Creativity-Intuition-Logic.

All work to tickle your amygdala.

Smile.

You have an Infinite Brain Machine right between your ears, at every moment.

Chapter 12
MORE CONVERSATIONS & AMYGDALA TICKLES

Vic Cooper
Automotive Custom Repair and Refinish Expert, Emissions Diagnostic Technical Expert

Anybody who thinks tickling your amygdala is just some New Age wimpy flower power girlie activity has missed the point completely. Vic Cooper has driven and repaired some of the leanest and meanest cars on the planet, and at any given time his garage is filled with a half dozen exotics and probably a dragster or two.

He knows exactly how important it is to tickle the best part of one's brain and he does it every day.

NS: "I'm interested in how you use your brain to problem solve. So can we start with a description of your professional background?"

VC: "I started building dune buggies when I was 15, bought my first Lotus Elan when I was twenty-something, which I still have. (Last time I checked he had six Loti in his garage, including a James Bond Esprit model).

I've been to all of the major American manufacturer's schools- Ford, General Motors, Chrysler. I've been working for the State of Colorado since 1978 and I've been doing the emissions diagnostics since 1981. In the early '80s when computer controlled vehicles first came, I was involved in writing the training materials for vocational schools on how to diagnose those problems. But cars have progressed a long way since then."

NS: "Besides the work at state emissions, describe what you do in your own business."

VC: "Back when I was a teenager, and I'm 56 now, I started doing fiberglass, and I still do that. I have four outside projects at home right now, and I regularly do race cars and custom work.

As an extension of what I do with emissions diagnostics, I'm also very good with electrical wiring, and as a side hobby I repair racing wiring harnesses for the Sports Car Club of America."

NS: "Can you name some of the kinds of exotic cars you've worked on?"

VC: "I've worked on Ferraris, De Tomaso Panteras, I've had Lamborghinis out at my house too, Lotus of course, a lot of exotics that you could think of. I've done ignition, carburetion, electrical, paint and body, and that's about as far as I go- I leave suspension to the shops. My daughter drag races, so I work on her two dragsters. I also work on sports motorcycles and the fuel injection and the fiberglass and carbon fiber parts for those."

NS: "You're a busy guy."

VC: "Well, the part I think you want to talk about is why someone fails at a diagnosis, and yet I can find the problem here.

The difference is that without being very intuitive about how a system works, someone else is stuck with a shop manual. The shop manual will give them instructions like this: If you have this fault, test this component- If it's good, test this- If you have this voltage, go here, etc. You're following a flow chart. You get down to the end of the flow chart and it will tell you what part to replace.

So if that's all you know about the system, you're relying on some engineer who wrote the repair instructions at the time that they built the car- and he doesn't know what kind of road salts are now being used, and the kind of corrosion you've got, and what kind of rodents are eating your wires.

Those people didn't think of every possible kind of failures. And if you're following their repair flow chart, and replacing the part they recommend doesn't fix a thing- you're going to end up here with me.

Our process is different. We *don't* look at the flow chart. We know how the system works, we know the logic behind the computer programming, and so with a handful of bits of information we can discover where the problem lies."

NS: "You have to be creative, because you face problems that are exceptional and unique, solving a new riddle. How important is it to learn from others?"

VC: "I think it's very important. It's important to learn what other people have had in terms of success, but also what failures people have had- what *didn't* work. So you don't repeat the same mistakes other people have made.

Here's one of my favorite examples involving a car that was bouncing around from one repair place to another, and nobody had a clue:

It was a Mazda 626, and it was failing emissions and two of the spark plugs kept fouling out. All the technicians knew that it was coming down to these two cylinders, numbers 2 and 3. But it was bouncing around all the dealer shops, and they were replacing fuel injectors, and replacing computers, and replacing wiring- they were replacing things based on what the shop manual was telling them what to do.

They kept hitting a dead end and the car still wasn't running right.

So the vehicle came to me here, and I didn't even look at the flow chart. When I looked at a wiring diagram- not a repair flow chart- I saw that the ground wire for 2 and 3 were indeed tied to the same pin on the computer, but also that there was also another component tied into those same two injectors, and that went off into the dashboard.

So I opened up the manual to the dashboard wiring, and it turned out that his vehicle has a fuel economy computer on top of the dash. It turned out that the owner had taken apart that unit and had modified a piece of the wiring.

But the story gets deeper than that.

So we cut the wire and corrected the fault, and the vehicle passed emissions immediately.

But the shops couldn't find that problem. And the owner had spent thousands and thousands of dollars trying to fix it. So, the owner took the dealership shops to court.

Everybody, including me, got to testify in front of a judge. Ultimately the judge ruled that the mechanics in the shop couldn't have known of any fault other than what the manufacturer's flow chart provided. Consequently, the shop hadn't done anything wrong."

NS: "Did the owner tell the dealership mechanics that he had messed with this other fuel economy computer?"

VC: "No. But it took us, here, two minutes to figure it out, where it had spent six months bouncing around these other places and they never could see the problem.

What I see this as, is a difference in *thinking*.

If you limit your thinking to what the flow chart says, you're going to be led down the wrong path."

NS: "What this seems to indicate to me in regards to problem solving, is that if you limit yourself to any kind conventional wisdom, you are going to run into difficulties. You need to see every problem as a fresh combination of elements. Especially if you're not solving the problem."

VC: "I think you have to understand as much as you can about how the system works that you're dealing with, but then you have to be *intuitive* about how you think beyond the written explanations of how it works."

NS: "In regards to that Mazda, you were looking at the car in a global way in how everything was connected- where as your failed predecessors weren't really observing…"

VC: "Right, they limited themselves to a printed diagnostic chart. They weren't looking at it with a fresh mind."

NS: "They weren't able to think *for themselves*."

VC: "Right. That type of intuitive thought is a daily thing around here, especially with the tuner cars [modified from stock] where people have altered various components, and they've added components like turbo-chargers- people can create problems when they begin moving things around.

We're constantly having to think outside the box, because everything that comes to me here is unique to begin with."

AMYGDALA TICKLE #26- "Almond Alternator"

Get some almonds and toss them on the dashboard of your car.

The next time someone honks their horn at you for any reason or if you get frustrated by a problem that doesn't want to go away-

Pop a couple of delicious almonds into your mouth and enjoy them.

This will tickle your amygdala forward instead of clicking backward into your reptile brain from another driver who is clicked backwards themselves. Intuit and find the right solution instead of spinning your wheels.

James Mullica
Screenwriter

Jim started out as a radio disc jockey in stations around the U.S. while in law school, and migrated into writing major network television scripts (Golden Girls, Empty Nest, Dear John). In subsequent years, he moved into writing feature film scripts as well as being involved in business management.

NS: "You've rubbed elbows with the rich and famous in Hollywood and other places..."

JM: "They're all human. The first time I met Angie Dickenson, I was with a friend who was fixing her TV. She was doing her wash, getting her T-shirts out of the laundry."

NS: (laughs) "Okay, but you've known a lot of people whom our culture looks up to as icons. People tend to think, 'Oh, if I was famous, if I had a big house, if I 'had it all' or 'if I only had a bit more', *then* I'd be happy. What's your conclusion about the relation of material wealth to success and happiness and fulfillment?"

JM: "People who have money are constantly chasing more. They're constantly looking over their shoulder because they're afraid somebody is going to take it from them. If you're chasing a dollar all the time, you'll never be happy, because you'll never get enough."

NS: "That reminds me, last night I told my wife Julia this story while we were walking through the neighborhood...

'The Buddha (before he was the Buddha) had been seeking enlightenment for a number of years. Slowly he had gotten rid of all of his possessions and all of the stuff he was attached to. He was down to this one little bag, the size of a small sock. Everything he had was in this one little bag.

But despite giving up so much, he had not yet transcended, he hadn't achieved Nirvana.

One day it finally dawned on him that he also had to give up this one last remaining little bag of just a few things. So he gave up this last parcel, and he became enlightened upon giving up everything.'

So after hearing this story, Julia said to me, "So how did he brush his teeth?"

JM: (laughs) "Well, from what I've seen, rich people don't derive their satisfaction from money. They may think they do, but I haven't seen it. One reason why I say this is because my father passed away a very wealthy man. Money didn't make him happy. I don't think he knew what made him happy.

He lived in a big mansion on the beach, and he once said to me, 'I was happier in that little two bedroom house we used to live in.'"

NS: "How about creativity- you're a creative guy. What do you do to spur your creative spark?"

JM: "I listen to music. I'll put on my headphones and get the music rolling. I believe in channeling- I believe that it comes from *out there*, that's its all existing out there some where and that I just need to tune into it and write it down.

I use music to open up a lot of things for me. I get ideas that I would never get if I wasn't listening to music. For instance, I'm now working on the script about the collapse of Wall Street, and I was stuck. The answer came in a lyric in a song I just happened to be listening to. One line in the song I just happened to be listening to gave me the answer I needed for the script."

NS: "Interesting- you keep your Right Brain happy with music while your Left Brain does the writing.

So, do you think there's any security in this world?"

JM: "None. Your security is within yourself. Money isn't, your wife isn't, your husband isn't, fame isn't. Believe me, all those things can be taken from you in an instant. It's the person who is comfortable in his own skin and finds security in his own self that has got it. You can't buy it. You can't buy happiness, you can't buy security. You either have it or you don't.

When I was younger, the idea of security to me was to always have a lot of money. When I was in high school I said to myself, 'I'm gonna' grow up and be a millionaire, and I'm gonna' be secure, and nobody is ever going to take it away from me, and I'll be happy.'

That all changed for me when my twin brother suddenly died, and I realized that everything can change in an instant. I realized that if I can just enjoy what I'm doing- that's my security.

Life isn't getting from birth to death. It's the voyage in between. If you enjoy that voyage on a daily basis and get the most out of that day- no matter what you do today, make sure it's worth it, because you've given up a day of your life for it."

NS: "How did you know what you wanted to do in life?"

JM: "I think if you get in touch with your inner self, you find out.

I used to listen to Larry King when I was twelve years old. I got a transistor radio for Christmas, and I would listen to his show with the radio under my pillow. One night he's interviewing this stripper, and that got my attention. I stayed up all night and then I tuned in the next night hoping that he would have another stripper on. He didn't of course.

Anyway, when I grew up, I ended up meeting Larry King through my father in-law who produced Jackie Gleason. King had a huge influence on my life and played an important part of getting me into radio. Incidentally, my father in-law also got me a scene in a Frank Sinatra movie "Tony Rome", but I screwed it up on camera twice, so I didn't go into acting. Afterwards Sinatra said to me, 'Don't give up your day job kid.'

But anyway, in my life, I've been lucky to meet all these people. I've seen the high society, the low society, I've seen how everybody has lived in between, and it seems like the people who don't have a dime are the happiest.

NS: "Buddha was right."

JM: "Yeah, but he lost his toothbrush."

NS: (laughs) "What do you do when you get stuck, like when writing and nothing is happening? How do you get the ball rolling again?

JM: "I stop and I walk away from it. I come back to it an hour later or a day later. I always find that if I sit and have a war with it, the creativity is not going to be there, its just going to be mush, words on paper that don't mean anything. If I'm stuck, I just put it down, and then out of the blue something will come into my head, and I know 'That's what I need to do!' It doesn't matter if it's a script or anything else."

NS: "What do you think about this idea that you have a switch in your brain that you can consciously tickle, that you can tickle yourself out of your reptilian brain and into the smart part of your brain? Do you think I'm totally insane or out of my mind?

JM: "Turn off your tape recorder. The interview is done. (smiles)

Ha ha. Okay, really, I think you can make your brain do a lot of things.

I see where people will not let their creativity shine through or cultivate their brain in order to allow their brain to allow them to receive. I don't think people even look at the brain as anything. It's a mistake."

NS: "Yeah. I don't think people even know that they have a brain- 'I tell people, 'You have a brain.' So they act surprised- 'What? I do? What's that? It's between my ears?? What's it for?!?'"

JM: (laughs) "I took a course on the brain in college. It was so interesting, because we learned how to go into this deep almost trance state, and consequently we found that we could recall things shown to us only for an instant on these flash cards that we shouldn't have any way to even consciously register or remember."

NS: "I used to have this thing that I would show to beginning piano students. They could instantly know where to find a chord on the piano with their eyes closed by trusting their intuition. You couldn't do it thinking about it consciously, but you could do it if you didn't think about it too much.

Most of what your brain does is not verbal, its not rational logic. Of course it's those things also, but most of what it does is something else. It relates to what you were saying about what happened in that brain course you took."

JM: "I have to tell you I was literally shocked when I saw that I could retune my brain so that I could recall words that I barely had time enough to read when I saw them flashed in front of me. Once we can calm ourselves down and relax, then all that information we're seeking just naturally opens up to us flows into our minds.

I'm a firm believer that there's an inner voice inside you that will supply you with answers. It's not going to allow you to walk in front of a bus. You just have to learn to listen to it."

AMYGDALA TICKLE #27- "Brain Treasure Chest"
There is an inexhaustible amount of information and research about the brain available to everyone. Go treasure hunting! Start exploring this Brain Treasure Chest of Information. Online. In Print. Medical Libraries, Public Libraries, Classes, Lectures. Draw your own informed conclusions from what you see, hear, read, and intuit from your own experience and sense.

Laurel Bouchier
Acupuncturist, Medical Herbalist, Reiki Master

LB: "A long time ago I decided I would live as though I had won the lottery, and didn't need money. It's amazing how well that works. That' how I got into all these things."

NS: "What's your definition of success?"

LB: "Being at peace and being happy. That's worth more than hundreds of thousands of dollars."

NS: "Instead of trying to achieve your goals, you acted as though you had completed your goal."

LB: "I think about what this teacher, Papaji, always said, 'Is *that* what you want, or do you want *what you think that is going to give you*?'

If you think it's the money that you want, what you have to think about is what do you think you're going to get- what's it going to get you if you have all the money? That's what you need to aim for."

NS: "Well, you are certainly one of the happiest persons I've ever known, that's for sure. How do you view pain and pleasure as far as guiding you towards your goals?"

LB: "Here's what I'm experimenting with- accepting pain if it comes to me, and working with it. But I'm also okay with having an easy, effortless experience. If we create our own experience, I'm for having mostly pleasure in my life and being okay with that."

NS: "I think a lot of people are brought up with the notion of 'No gain without pain.' Do you think about that?"

LB: "I think that's total baloney. I think we are coming out of the dark ages. You can relax.

Hey, I think of myself as immortal. Some things you do take hours, some things take days. Other things take weeks and years to get. But other things take lifetimes to learn. Don't worry about it. You don't have to get everything done by the time you're eighty. I might have to switch out bodies.

So, thinking that way relaxes me. It doesn't mean I'm not going to work on myself, but some things are going to take lifetimes to get right, so just chill."

NS: "How are you guided by your higher intuition on a day to day basis?"

LB: "I use it all the time. For example, just like buying fruit. One has a little more sparkle to it, a little aura around it, like 'Pick me!' and I go, 'Well, that's the one!' Or if I have four or five ways to go somewhere, I'll just have this little picture in my third eye of what street I need to be on and which way I need to turn."

NS: "It sounds like that for you that this tuning into inner feedback is something that goes on with you all the time..."

LB: "Constantly. I trust it more than any computer on Earth or anybody else. It anybody else wanted to tell me what to do, I would have to check with myself first. I trust that more than anything. I think intuition is way undervalued in our society."

NS: "Can you elaborate some more on that feedback mechanism?"

LB: "A long time ago I learned that whenever you're trying to decide something, take a vote between your mind, and your heart and your gut. Like, what is your stomach saying? To me, your gut was your intuition.

Intellectually you can spell it out with your logical mind, like, how much is this going to cost, and what about the interest on my charge card, and do I really need it. But then my gut is telling me, 'I really want it! I have to have it!' Or, I don't *really* need that thing. Take a vote. You have to have foresight-you can't just be stuck in that very moment."

NS: "You have to be able to go down the time line."

LB: "Exactly."

NS: "You learn to look at the global picture rather than just your immediate reward. And then after a while immediate reward changes.

For example, a teenager might not at first think about anything except what is right in front of his nose, so they might act impulsively, 'This is going to feel good right now! So I'm gonna' do it!'"

LB: "So they don't think ahead..."

NS: "But as you mature and use your frontal lobes, you incorporate the long term effects of something into how you perceives something momentarily. So then, you can no longer binge drink..."

LB: "Because you're going to pay for it. But I think we're now back to what Papaji said, 'Do you want *that*, or what you think it's going to give you?' Because you can have what you think it's going to give you *right now*.

Nothing has to happen to get IT. You can have what you *really* want-right at this moment, you don't need anything else to get IT.

It's *here*. We've already got it all. We don't need to do anything. Do you want the money or do you want what you think it's going to get you? Happiness? What do you think all these things are going to get you? Freedom? Love? What is it?"

Like the family that was poverty stricken, looking everywhere for jobs, and selling things. But they had a pound of gold underneath the floorboard in the kitchen. All they had to do was to lift up the boards. There it was. Hahah!"

AMYGDALA TICKLE #28- "The Medium is The Massage"

Get or give yourself a massage. Learn about foot massage, or buy an inexpensive electric foot massager. Tickle your brain while you tickle your toes.

Kyle Ridgeway
Physical Therapist, DPT (Doctor of Physical Therapy, Neuroscience)

NS: "You majored in neuroscience?"

KR: "Yep."

NS: "The amygdala and the limbic system are a reward/punishment system within the brain..."

KR: "Absolutely."

NS: "When we encounter something that is a disadvantage or a threat to our long term survival we are given negative emotional feedback by this system..."

KR: "Absolutely."

NS: "When we do something that enhances our long term survival or quality of life, the brain signals this to us..."

KR: "With *reward*. But people are typically very bad at discriminating between long and short term reward, because of "loss aversion". Neuroscientist Antonio Damasio did some interesting experiments to show how loss aversion affects our decision making. This is the tendency to avoid a rational decision because you've just lost something. You have to be able to engage logic [left frontal lobes] to overcome that.

NS: "So what you are saying that we are limited by this loss aversion that takes place in the reactive processes of our brain, but we can overcome this with higher cerebral processes that occur in the frontal lobes?"

KR: "Absolutely. The higher frontal lobe task inhibits the inherently quick, very emotionally based incomplete information.

Damasio also discovered in a particular gambling task experiment that people began to use the correct winning strategy even before they could explain what they were doing."

NS: "So what you're saying is that there is also this intuitive process [right frontal lobes] that allows us to make good rational decisions?"

KR: "Yes, even before we can actually explain what we're doing. The funny thing was when he started measuring people's nervous system responses with galvanic skin response, the subject could detect the difference between choosing a high risk decision and a low risk decision, even before they were consciously aware of the difference."

NS: "So, when you talk about frontal lobes processes, it's not necessarily a verbal "rational" process. That's only one small slice of what the frontal lobes do."

KR: "Absolutely."

NS: "Thinking about what you said earlier, are you suggesting that if we get caught up in negative emotions, it ultimately leads to bad decision making and bad problem solving?"

KR: "Absolutely. But the flip can also be true. If we are guided only by positive emotions, we can also go down a slippery path as well. There are a lot of things that initially feel very good but turn out to be very harmful in the long run, including things that can lead to an addiction."

NS: "So then, if we are only concerned with short term negative emotions or short term positive emotions, we're screwed either way. But if we can integrate perception of long term consequences [frontal lobes logic/intuition], that's to our distinct advantage-"

KR: "Absolutely, I would agree with that one-hundred percent. The difficulty is in imparting this idea to the general population."

NS: "I would suggest that one can consciously cause a direct increase in frontal lobes processes simply by "imaging" that within the brain.

Imagine you have a feather, and you tickle the front of each amygdala. In doing so, you cause energy to increase in the frontal lobes, simply because you're engaging your imagination. That, in essence, is like putting a little gas in your car engine's carburetor- you're priming your frontal lobes to start up. You think, 'Oh yeah, I can engage my frontal lobes instead of just reacting to this situation.'

But also, using the imagination to remind a person to use their frontal lobes actually also causes a direct increase in frontal lobes activity by its very nature."

KR: "I believe that the idea of tickling the amygdala is definitely a helpful thing. Just the thought of the feather engages your imagination.

In my work, in chronic and persistent pain, we first use "graded motor imagery" to use the imagination in reducing pain before we use the physical system."

NS: "So you're engaging the imagination first?"

KR: "Exactly. Graded motor imagery."

NS: "What you're doing is engaging the higher brain functions to overcome the limitations of the primitive brain."

KR: "Right, to retrain all these systems to eliminate pain from a top-down manner. And we engage these sub-conscious systems as well, so it's very interesting."

NS: "Is pain essentially a product of primitive brain processes?"

KR: "That's probably accurate. But pain can be affected by higher processes, both for good and for bad. For example, if you bump your elbow, it'll hurt, and you probably will get over it shortly. But if you begin to imagine with a higher level process [frontal lobes imagination] that 'Oh no, what if I broke my elbow? What if this is something serious? What if I lose my arm?!' and you start down this road of catastrophe, the pain is probably going to get worse."

NS: "But the cure for that is also imagination- that's why we talk about "*tickling* your amygdala" by engaging a positive imaginative process that feels good."

KR: "Yeah, you don't want to poke it. That's what I talk about a lot with my patients and have them conceptualize their pain in a non-threatening manner. What is pain? You don't have 'pain' receptors, you have *danger* receptors that interpret signals from your body. When you brain interprets enough danger it gives you pain to get you to do something about it."

NS: "Well, so pain is the signal to get you to stop doing something that's dangerous, it's signaling to you that 'this is not good for you'. What do you see the role then, of pleasure? What are those feel good emotions telling the individual?"

KR: "At a basic standpoint you would have to say that if something is pleasurable and feels good that it either enhances survival or health. But we've evolved to the point in society to where that primitive level of short-term reward is not enough."

NS: "You mean that we have to work at a higher level than a squirrel who sees a peanut butter sandwich in a squirrel trap and thinks, 'Oh, that smells good, that tastes good.' But of course, that's the end of the squirrel.

KR: "I would say that we're worse- the squirrel doesn't have that ability to say, 'Wait a second... let me think about what might happen if I grab that sandwich...'

If we don't use our frontal lobes then we're worse off than those lower animals because we're not reaching our potential. We have much bigger frontal lobes than a squirrel. Taking away our frontal lobes is much worse for us. You can see the loss of frontal lobes engagement in our world and what results."

NS: "Isn't that what advertising is about? That we're culturally conditioned not to turn on our frontal lobes?"

KR: "Exactly. Our educational system is constructed so it teaches fact and regurgitation. We don't teach *thinking*. We don't teach the philosophy of science as a foundation to help you to grow all the time.

A common saying at medical field graduation is, '50% of what we've taught you will no longer be valid in ten years.'

So, in other words, the most important thing is not *what* you've learned, but *the ability to think.* Unfortunately we still fail at that, even at the doctorate level of education in this country."

NS: "Isn't it true that the reactive parts of your brain don't have the ability to learn any new information? That this primitive part of your brain is just running off of genetic and cultural pre-programming?"

KR: "Well, it can't learn anything in the conscious sense, but it can still be changed- It can still habituate to new inputs, it can still sensitize to other inputs..."

NS: "But it can't anticipate. It can't predict."

KR: "But maybe we can teach it from inputs from the periphery and inputs from the frontal lobes back down."

NS: "Historically that's a problem with humankind. Humans get used to solving problems in the same old way, but the environment is constantly changing. When a new problem comes along and we try to solve it with old solutions, we're screwed."

KR: "My favorite Einstein quote is, 'To do the same thing over and over again and expect new results is the definition of stupidity.' And yet as humans, we do that consistently."

NS: "The objective here is to recognize a rudimentary tool, a basic thing so people can instantly move in the right direction. One that'll work is "Tickle Your Amygdala", however each individual chooses to do it in their own way.

KR: "That would be great for teenagers. It can't get much simpler than that."

(The conversation continued the following day)

KR: "I've been thinking about tickling the amygdala all day yesterday. It was an interesting thing. It's definitely a helpful thing for me, definitely."

NS: "So are you doing the exercise where you're visualizing that feather?"

KR: "Yeah, there were times when I did it just briefly, and then there were a couple of times where I just sat there and pictured where the amygdala was anatomically, and I tickled it with a feather."

NS: "I get a wide range of feedback- most of the time it's positive, where they get mood elevation, or they solve a problem. Sometimes people get a strongly positive and euphoric reaction. Sometimes I get the reaction where people report nothing happens.

KR: "Well, that's just their *perception* that nothing happens. Who knows what really happened? It still may have been of benefit even if they don't consciously perceive it at that moment."

NS: "I tell people that it's very helpful to keep a log- you survey yourself. You make a graph out over a period of time how you view your daily experiences; physically, emotionally, intellectually, and maybe even spiritually. You rate from minus ten (negative) to plus ten (positive). You do this once a day over a period of time as you practice tickling your amygdala, so you can judge the effects in an objective manner. You don't take my word for it, or anyone else. You make your own decision whether this is nonsense or not."

KR: "The thing that I was really struck by was what we discussed yesterday- The engagement of the imagination. Just the thought of that feather really engages your imagination, it's instant. It is so simple, but it's an instantly effective means for engaging your frontal lobes and higher processes."

NS: "Like I said earlier, it's like priming a car engine to get it started, you put a little gas right at the carburetor, and the engine starts up."

Now, I realize, some people can't imagine a feather inside their head- like this yoga teacher I know. So I said to her, "You go get a real feather and put it in your pocket. And once in a while, you take it out and you actually tickle your forehead, and do it with your eyes open. Do it in a mirror for a while. Then after a while, close your eyes and you'll probably be able to see the feather and do it without holding a real feather to their head.

But even so, people can just verbalize it and say, "I've got a feather, and I'm tickling my amygdala, and I'm causing more energy to flow in my frontal lobes." Just say it with spoken or silent verbalization. This will engage various parts of the frontal lobes, and also abstract conceptualization in the pre-frontal cortex.

Anyway, what happened when you tickled your amygdala?"

KR: "I could definitely feel myself becoming much more frontal lobes oriented. I felt like I could definitely take more control, if not just of my thoughts but at least control the influence of those lower levels of my brain of over my thought processes."

NS: "You felt like you were moving into more advanced processes of your brain?"

KR: "Absolutely. This could be a real groundbreaking thought process for physicians and health care providers.

231

One of the things that's really important in clinical decision making whether its medicine or any other health care profession is whether you're making some kind of reflection in meta-cognition [thinking about thinking] so you can see where you're falling into cognitive traps in your decision making, whether it's diagnosis, whether it's interaction, whether it's prejudices.

Even when you try your hardest, we all have biases and pre-judgments of the person that's sitting in front of us and those can cloud our actual higher level processes of the data or the information we're getting from this patient and everything else to make the right decision.

This simple concept can be a grounding framework of "Wait, wait... let me tickle my amygdala here, I've got this gut reaction..."

There's confirmation bias- we give more weight to positive results than we give to negative results..."

NS: "You think that's good?"

KR: "It's a bad thing in medicine- 'This stuff was negative, and goes against what I was thinking, so I don't give it as much weight as what confirms my hypothesis...' The other thing that can happen is a diagnosis of convenience, for example, if I'm in an emergency room and I've just seen five people who come in with pneumonia, I'm going to be more apt to see the sixth person with pneumonia as well."

NS: "So how does tickling your amygdala help out?"

KR: "So if I'm tickling my amygdala, first of all it makes my pause for a second to engage my imagination, Then second, it's a cue to ground my thinking- "Let me look at this logically, let me look at this information before I jump to any quick lower level automatic conclusions which could be totally wrong'.

NS: "So by turning on frontal lobes logic and intuition, you just don't react..."

KR: "Exactly. The other thing that can help is the interaction with the patient. Because as a doctor I probably have all these other things going on, and by tickling my amygdala it puts me right here with this patient, and it makes me cooperate and interact with that patient.

In graduate health science programs and medical school they're starting to talk a lot more about emotional intelligence. Being a good physician is not just about intellectual prowess or good diagnosis- its also largely about emotions as well and how you connect with that patient, how you are able to educate, communicate, and empathize with them. That has a very profound effect.

Incidentally, the chance of being sued for malpractice has statistically nothing to do with how many errors you make. It's based almost solely on whether or not your patients like you. If you are a jerk and your patient doesn't like you, you are much more likely to be sued for malpractice. The data for this is virtually incontrovertible.

I think that by tickling the amygdala may help us engage this emotional intelligence, which does take focused frontal lobes thinking. It is hard to get beyond your biases, and prejudgments, and empathize with every patient you see, all day long. Tickling your amygdala is a time out that makes you pause to engage that higher level thinking."

NS: "Nicely put."

KR: "You can quote me." (laughs)

AMYGDALA TICKLE #29- "Tickle In Your Pocket"

The only real trick in tickling your amygdala is remembering to do it- So, get a REAL feather and stick it in your shirt pocket or pin it to your lapel or shirt sleeve.

Throughout the day, look at it and remember to use the best part of your brain. You might even grab it and brush it against your forehead or temple for a tactile reminder.

Britt Severson, M.D.
Physician, Family Medicine

NS: "When you were a medical student, in many ways you were delaying gratification- which is something you need your frontal lobes for. You understand that the jackpot is a little further down the road."

BS: "When you first start medical school, for the amount of work you do, you get about zero gratification. You work your butt off and you don't even see patients until your third year, and you're still low man on the totem poll. At the beginning, I had zero enjoyment and I just wanted to quit. But if you manage to talk to someone who's been through it, it helps to drive you on."

NS: "So, you're coping with the inevitable stress inherent in the medical education system?"

BS: "It's the light at the end of the tunnel knowing that I could build my practice the way I wanted when I'm done. But one thing that I did when I wanted to quit was to take a year off and finish my master's degree in public health. That helped me to refocus on what I felt was important to me instead of being so caught up in just endlessly stressing out in medical school. I came back being refreshed from being able to hang out with people who are not in medical school."

NS: "What makes a good doctor?"

BS: "The funny thing is, they breed doctors to have the best grades and have the best research, and be the smartest to into medical school. But that does not make a good doctor in any way. What you really need to get through medical school is simply the willingness and ability to work hard. But when you get out, the people who make the best doctors are the ones with a well rounded background and have a higher *emotional* IQ- those who can feel empathy and can listen to patients."

NS: "I've seen a lot of evidence that attitude plays a huge role in healing. In some cultures, it's almost everything."

BS: "I think the effect of the mind on wellness and disease is huge. Studies have been done to show that things like meditation and mind control have a dramatic effect on the immune system. In medicine, you can look at things from many angles. A big part of the picture is definitely how you sell it as the doctor, and as to yourself as the patient."

AMYGDALA TICKLE #30- "Laughter Is The Best Medicine"

Next time you're not feeling up to par, do what smart people do- Start laughing.

It's been shown repeatedly that a half hour of smiling, chuckling, and flat out guffaws has as much pain killing healing power as many potent prescription medicines- at far less cost.

Bobby Reginelli
Journalist, Entrepreneur

My friendship with Bobby began when I showed up at his parent's house to begin giving the then fifteen year-old his first guitar lessons. We continued as teacher and student through his high school years until his departure for college, but we kept in touch all through to his graduation where upon he earned a degree in journalism and went to writing a local newspaper column.

During his college years his guitar playing developed further into a keen and deep interest in music. He eventually managed the college radio station and starting his own foray into the music industry. He first worked at local concert venues and then created his self-run business of concert and sports ticket sales. Within nine months he had made enough money to live off of and travel the globe for the succeeding two years. We began our interview at a local coffee shop.

BR: [Bobby pointed to a thousand-dollar carbon fiber road bike that he rode that morning. Our Brain Radar together had it found for sale on the very day he had called me for help in finding a new bike to buy. It cost him a mere fifty dollars for the frame, and then later, just another seventy five for the wheels.]

"Remember when we got this? I can't believe we found some wheels to exactly match the color I painted the frame."

NS: "Brain Radar strikes again. Why do you think that keeps happening to you?"

BR: "I try to stay open to anything. My parents always told me, 'You will do many things in life.'"

NS: "You had amazing success in your ticket business. How did that come about?"

BR: "I knew what music was hot from my job managing the radio station. So I'd get ten of my friends and we'd swarm into the school library, each person grabbing three computers. We'd start at eight in the morning and get tickets from all over the country. By the afternoon we'd have tickets to sell for concerts from Boston to Honolulu."

NS: "So this was a natural extension of what you liked to do anyway: music. Sounds like your frontal lobes kissed your amygdala."

BR: "In nine months we grossed three-quarters of a million dollars. After expenses and taxes my partner and I pocketed seventy-thousand dollars each.

Starting out, fun was a big part of how we ran things. Sometimes getting somebody a ticket- like twenty minutes before some sports match- was like being in some outrageous movie and all kinds of crazy stuff would happen.

I'd get on the phone to some guy in New York and he'd say, 'Yo, this is Louie. I'm lookin' for three in the Diamond Club, and I'm lookin' for 'em right now. I've got five-hundred to spend, so you get me some tickets...' and he'd hang up.

So, we'd go out and find some tickets. But I'd have to get on Google maps because I had no idea where anything in New York was, and the guy who had the tickets I would be selling to Louie would tell me, 'Alright, I've got a guy who's got all my tickets over at Shelly's Candy Market. That's Shelly's Candy market at 14th and Providence. Okay? You tell your customer to go over to 15th and Providence and go to behind at the counter and ask for Jerry.' It was insane. It was like a crummy Mafia B-movie plot, but about tickets.

So, I'd be sitting there with my finger on a street map of Manhattan trying to coordinate this thing like a secret military operation. I'd get back on the phone and call Louie and tell him, 'Okay, listen, we got your tickets. You're gonna' hop on the J subway and head uptown. You're gonna' get off the train at 15th and cross over the train tracks and get to the candy shop, go inside and ask for Jerry...' (laughs)

That kind of thing when it was at the last minute was fun. We'd make up competitions between ourselves to sell tickets, and the loser would have to buy lunch. Yeah, we definitely tried to keep it fun."

NS: "So why did you quit doing it?"

BR: "After nine months, we had enough. The ticket business, and I don't mean that we were, is rife with corruption and unfairness through and through. It's a business that makes people angry. No ticket should cost what people were being asked to pay. People were angry, sometimes justifiably so. But having angry people in your life is horrible. It was all filtering down to us. We were in attack and counter-attack mode. What started out as enjoyment eventually capsized over the nature of the beast. It became unhealthy, and not worth it any more, no matter how much we made.'

There was also the concept of value. I asked myself, 'Am I adding value to the world?' There's an argument that we were providing a needed service, especially in the case of popular shows where someone wouldn't have a chance to attend unless they came to us. But in the end we just found ourselves hanging on the coattails of someone's popularity, and it wasn't fulfilling. We were running arbitrage on a commodity that was limited- seats at a show.

When you're charged with adrenaline, everything but basic functions shuts off. And you and I are not the type of people who operate well at just basic functions. We need higher functions to make our mark.

Besides all that, in our business, we were at a very distant point from control. We had no control over whether the product- the ticket- would be a success or not. You just never knew what was going to happen, and somebody else had the control. It sucked."

NS: "How would you rate that business on a scale from 1 to 10? Minus 10 being suicidal and plus ten being Nirvana?"

BR: "At the beginning, +7. At the end, the emotional feedback was down to just +2. The excitement wore off. The final straw that was that we lost all the balance in our lives. That was the real thing that did it in.

At the end of the day we didn't have the knowledge or the perspective that you need to balance your work-life out with the rest of your life. So you make a million dollars, what good is it by itself?

Ultimately, I didn't enjoy it. It didn't make me feel as good as I knew I was capable of feeling. It was something that I knew inside at some level, 'You can do much better than this.'"

NS: "So your amygdala and your emotional feedback was telling you, 'Not this way.' Money couldn't fill the cup."

BR: "Right. My partner and I also had this T-shirt business going, and one thing we noticed was that by contrast everybody who ever got one of our T-shirts was just always so stoked about it. It was win-win. We didn't make as much money that way, but we had that to compare the ticket business to, and it opened our eyes.

What I learned when we stopped and I started traveling was that I wanted to be in a business where what I did made your life better, where everyone involved left with a great feeling about the transaction. That was something I saw everywhere I went, no matter how much money exchanged hands.

Every business that I do from now on, I build that in. Because if you can't do anything sustainably, then you're not going to succeed- you'll be on a timed fuse. You got to be able to win the race, not just really run fast for a mile."

NS: "You've traveled all over the world now. What have you learned from that?"

BR: "Here it is: What's important, is people. People in your life are important. I'm trying to make that a priority. I want to enrich people's lives by my interaction with them. That's what I saw when I traveled: Community is really important."

NS: "You've been into horticulture this past couple of years. How does that fit into the picture?"

BR: "One of my friends told me this, 'If everybody would just tend their own gardens, the world would be a much more stable place.' The way I see it, we're just put on this earth the tend gardens."

NS: "That's good. I mean, what better garden do you have than what is growing between your own two ears?"

BR: (laughs) "Perfect."

AMYGDALA TICKLE #31- "Neural Nature Nurture"

Get a new plant. Place it on your desk. Take care of it. Watch it grow. It's like the neural tendrils inside your brain.

Look at it and think about all the growth of new brain connections in your brain- always connecting in new ways, always growing- provided you continue to tickle your amygdala!

Chuck Schneider
Jazz Musician, Teacher

Chuck Schneider is a virtuoso on saxophone, clarinet, and flute, in addition to being a composer. Among other accomplishments, he has spent time on the road with famed tenor sax legend Stan Getz, he was additionally recruited by Clark Terry for his touring group, and he has played in countless other small and large groups across the continent. Chuck has been a music teacher for decades and is universally adored by his students not only for his boundless knowledge but also for the manner in which he gives instruction.

NS: "How did you get started in music?"

CS: "I was just an average B flat kid, and that's why I probably ended up playing the Bb clarinet and Bb saxophone." (laughs)

I joined the Highlander Boys youth group at the age of eight, picking up the clarinet. In that group, you could either march flipping guns around or join the band. I picked the band just 'cause a couple of my friends did.

My parents never pushed me, but they supported me wherever I wanted to go. I was always first chair, because I was driven- because I loved it. I loved playing and I loved listening to music.

But I never liked that egotistical competition in music. Those are the wrong reasons to be in music as far as I'm concerned. I get my satisfaction from joy of the music and the art form.

That part of it hasn't changed- I do it for the enjoyment not because I get a bunch of applause or not. I've played some crappy solos in my life and they just go crazy, and then I've played some great stuff and the audience doesn't even care."

NS: "But I also know you're interested in a lot of things- philosophy and such…"

CS: "At the University of Denver I wanted to go into psychology. But because my name starts with "S", by the time they got to the S's to pick their classes, all the psychology courses were filled and they had nothing left. So I had to take anthropology and poetry. They ended up being my favorite classes outside of music! I studied math, philosophy, poetry, anthropology and, all those things you couldn't make a living at. (laughs)

I switched majors a bunch of times, but suddenly my father passed away just like Bang!, and that just blew my world away. So I was going to drop out, get a motorcycle, and head south with a friend of mine just like Easy Rider.

I had been laying around at night listening to the radio, listening to Eddie Harris and Jimi Hendrix, messing around with the guitar. I couldn't explain why, but suddenly I just felt through my mind that *I had to be in music* rather than waste my life on a motorcycle. It was just totally *intuition*, it was saying 'As stupid as it looks, this is the thing that you've always really enjoyed.'

So I went back and got an audition with the DU School of Music. The next thing I know I'm the third chair clarinetist in the university band- not bad for having absolutely no real education in music.

It was just something that pulled me back to music. There was no rational certainty in it. How could I possibly make a living and raise a family teaching music lessons and playing gigs? I couldn't have imagined, there was no way I would know how it could be done. It was just a feeling of that's what I should do, *no matter what*. As it turned out, I actually got a half-tuition scholarship, an even better situation than before as a 'serious' student."

NS: "Did you ever have one of those 'brain pops', you know, this heightened sense when suddenly 'Eureka!' and all the pieces fit together?"

CS: "As I was going along, once in a while there were these magic moments. I didn't know what I was doing, but I might be told to take an improvised solo at some gig, and suddenly everybody would sit up and go 'Wow- what was that?!' and I'd think, 'Gee, well yeah, maybe I ought to really keep at this jazz thing!'"

NS: (laughs) "You're known as a really great guy to work with, as opposed to being a temperamental know-it-all and prima donna. How's that affect your students? You know that saying, 'Nice guys finish last.'"

CS: "I don't believe that. I've had a few students that left me because they didn't feel like I was drill sergeant oriented enough, but that's been very rare. I always feel that people respond when they feel good about themselves, when they're encouraged and there's no combat going on.

Basically induce that state of relaxation and whatever it is that turns somebody on to their thing. Cooperation is the thing for me and my students. Especially for yourself, when you cooperate with what you know is right *for you*."

Marilyn Auer
Publisher/Editor-in-Chief, Bloomsbury Review National Book Review Magazine

NS: "Now I do have these special interview glasses that I've found helps me to do a better interview... [-hands Marilyn a pair of laser etched light diffraction special effects 'Rainbow' glasses] Perhaps you would life to try them on..."

MA: "Let me see..."

NS: "This will help this become an Interview From Another Dimension... you see?"

MA: [puts on glasses] "...Oh yeah. Actually... this is exactly like an ocular migraine..."

NS: "Oh. Alright. Never mind."

MA: (laughs) "Hahaha!"

NS: (aside) "Just trying to introduce a little levity into the conversation. (Ahem.) So anyway, I decided that all of the problems in the world for which humans are responsible basically comes down to one thing, which is that people don't *think*."

MA: "You've noticed."

NS: "So, I'm going around and talking to people that I've recognized *do* know how to think in one manner or another, and perhaps I can share some of these ideas with people who are not thinking at all or thinking far too little. And perhaps in this way we can all together save humanity from total implosion."

MA: (laughs) "Oh, okay. Well I think it's not only that people don't have information, or that their sources of information are questionable, but that they don't think of things critically. So when an idea is put out there you think, 'Huh?' and you have to track it, and you have to cross track it, and you have to ask other people. So you have to have a lot of sources of information."

NS: "How do you see society in general, in terms of how people get their information? How do you think most people get their information?"

MA: "I think they watch bad news that is making mis-statements, and inflated statements."

NS: "Like the news that has turned into Cherry Kool-Aid?"

MA: "As Paul Conly said, it's 'News You Can Dance To'."

NS: "I was thinking about this on the way over here- people are being sold this idea, 'Don't think. This device will think for you.' Isn't this what this popular new cell phone is all about- 'You don't have to think any more.'"

MA: "And Google Plus? 'Here, let us take you where we think you want to go.' But also I think that there's so much information out there that people don't think they have the time to digest it, so they just want someone's opinion. Someone they trust- only those people are often not very clear thinking people either."

NS: "The people supplying the information..."

MA: "Yes."

NS: "You have a lot of contacts in the work that you do, a lot of authors and so on. Have you made a connection between people who are successful or unsuccessful at their work as writers, and the mind set that goes with each?"

MA: "I think some people feel that they need attention and acknowledgement and recognition. And so they seek it in a way that I think is obstructive for them."

NS: "Apart from the value of their work in itself?"

MA: "Right. And then there are other people who are so enchanted with their gift, that this is the source of their focus. For instance, there is a fellow named Gary Nabhan who is an ethno-botanist who does a lot of work with food. He started the Native Seed Search down in Tucson. And he is color blind, but has done breakthrough things. For example, he has taken botanists and color blind botanists out into the field for plant recognition to study cross-pollination. What he discovered was that the color-blind people could find forty-percent more of what they were looking for because they were seeing differently.

Now this guy, who is probably introverted, has done all of this breakthrough research with a gift of inhibited color perception that was not presented to him as any kind of a gift when he was a child. But in the end, now he's changing medicine."

NS: "So what you're saying is that here's a guy who's not after publicity, and not after fame, and probably not after fortune- but he's come up with these breakthrough things as a result of being primarily focused on his task..."

MA: "And exploring his own mind-"

NS: "As opposed to somebody who is working for the reward-"

MA: "Which is very fickle."

NS: "You don't really have control over that."

MA: "And you set yourself up because they love you, then they hate you because of whatever is the fad at the moment."

NS: "Do you see a lot of that in the literary circles and the book business?"

MA: "I do."

NS: "The most popular books are not necessarily the best books- Is that a good observation?"

MA: "Well, that's our approach. That there are really good books that people never hear about because they're not promoted. If there's a book that comes out from the University of Arizona that's really breakthrough in its own way, it's not really going to be the next... whoever's on the Top Ten right now."

NS: "Is that what you see as a part of your job, to help people learn about these good books that don't have a whole page spread in the New York Times?"

MA: "Yes, that's exactly our point."

NS: "So your emphasis is on quality rather than on commercial success or..."

MA: "But sometimes they go hand in hand, and that's good."

NS: "It seems like the trend is towards less reading in our society."

MA: "Well, I hear a lot about that, and I read about that- but among the people I know, that's not the case. So maybe my circle is too small." (laughs)

NS: "But seriously, why do you think people are content not to educate themselves, and read and learn, and get information from a variety of sources?"

MA: "I think people are overwhelmed- by the noise and by the flashing lights and everything. I think neurologically we're challenged."

NS: "Distractions, too many distractions."

MA: "Yeah. T.S. Elliot said that, 'We're distracted from distraction by distraction.'"

NS: "What is unique about books that you don't think we can get from the World Wide Web and television?"

MA: "I think books are a lot more personal. A book is an invitation to spent time with one person whose ideas you're reading."

NS: "Is that because you can flip from one thing to another so quickly on the web? If you get bored too quickly with one web site, you just click and you're somewhere else-"

MA: "Oh yeah. But a book, you've agreed to go to that world."

NS: "Like talking to a person."

MA: "Yes. That's my feeling."

NS: "You go into more depth than is the tendency online, certainly more so than in a cell phone text message."

MA: "Right. And for me, reading a book is a very quiet thing to be doing. It shuts out the noise."

NS: "I like the fact that I can take my book and go sit in the botanic gardens outside. And the battery won't wear out..."

MA: (laughs)

NS: "Also, I do know, since I publish my books both electronically and in print, *there's a difference*. A book just *feels* different."

MA: "It's tactile. I don't really think its one versus the other, I think they're just both here. One of my brothers is very involved in technology, and he loves to buy all of the new things, but he's also very supportive of all the arts. He has a Kindle and an iPad and everything, and then he buys printed books to read. I think they just co-exist.

Books are cheap by comparison. That's what libraries are for too, those that are still dealing with books, and I think there are a few of those left.

There's an interesting book that I recently got from the library called, "How To Think Like A Neanderthal". Apparently the Neanderthal brain was proportionately larger than our [Homo sapien's] brain."

NS: "Was it?"

MA: "Yes."

NS: "I suppose I should know that as a 'brain guy'."

MA: "You should."

NS: "I'm embarrassed to admit that I would have flunked that test question."

MA: (laughs) "There was an article in the New York Times about some of the breakthrough discoveries studying Neanderthal remains. They were also apparently quite communal and more social than we previously thought."

NS: "They may have very well been smarter than us as well.

A long time ago I had a discussion with friend who was a surgeon about this sort of thing: How did the frontal lobes evolve in the human skull? It seems as though we don't use our frontal lobes as much today as did in pre-historic times.

Those ancient people didn't have the comforts, and pre-digested frozen foods, and they couldn't just go to 7-11 to pick up ice cream, or go to Target to get shoes. So they had to be much more creative in surviving than modern people are.

Those individuals who survived by having more neurons to use, passed along to their offspring the genes and advantages of having a bigger brain. So, we developed a larger pre-frontal cortex, which was used back then a lot more than your next door neighbor uses his pre-frontal cortex today."

MA: "Also I think that pattern recognition was better. I often think that noise level obliterates perception now. Didn't Aldous Huxley suggest that we've narrowed down this broadband of perception to a very narrow band?"

NS: "We filter out stuff, because otherwise we'd be overwhelmed. But to filter out so much desensitizes us to some degree. So you end up losing some of your perceptions along the way."

MA: "It's like you lose a range of your hearing."

NS: "You build up a tolerance-"

MA: "-For chaos."

NS: "If you're at a rock concert and the music is loud, after a while you can't hear anyone talk at a normal volume anymore. I think that's why people go camping, or go to a Zen retreat, or some place like that so they can re-sensitize themselves instead of being de-sensitized."

MA: "I'm thinking about a time that I spent at the border of New Mexico and Arizona on the Tohono O'odham reservation. It was so quiet there; it was astonishing how restful it was to not constantly hear something. It was the best sleep I have ever had. Just being awake- there wasn't anything!

But that brings to mind, that I think the senses are far more connected than people realize. If we're seeing something, we're also feeling something like a tactile thing. Apparently in autism, many people are mono-sensory. So, if they're using one sense, they don't have access to the other senses. My nephew is like that, if he's looking at something, he can't hear anything."

NS: "I think what you're pointing out is that the senses inter-mingle. So, to connect what we were talking about a moment ago, if for example we hear a loud noise, it's going to have an effect on our thinking, or our hearing and visual senses."

MA: "Wouldn't that trigger our reptile brain, if it's a really loud noise?"

NS: "That's right. If you are stressed out by loud noises, it would automatically cut off your frontal lobes- you couldn't add three plus seven. And wherever you go, stuff is blaring at you in one form or another, and it's dumbing us down."

MA: "And contributing to ADD, and all that other stuff."

NS: "So, everybody go home now and smash your TV..."

MA: (laughs) "Or turn it off. Or when someone says something, don't just believe it. Use your brain."

AMYGDALA TICKLE #33- "Ear Plug Nirvana"

Get some of those nice comfy foam ear plugs. Keep them handy for when the cubicle next door starts to frazzle your fun.

A good meditation is to find a quiet spot, insert your portable isolation booth plug-ins, and listen to the very, very, very quiet sound inside your own head, from the ear plugs inward. Just listen. Nothing else.

Kent Miller
Attorney

Kent Miller's encyclopedic text books on law can be found in every Colorado practicing attorney's library and have become standard reference volumes for personal injury cases. During his forty-two years of law practice, he has received numerous professional awards including "Exhibiting Highest Standards of Competency, Ethics and Professionalism", *Colorado Defense Lawyers Association* (2007), and Top 100 Trial Lawyers in Colorado, *American Trial Lawyers Association* (2010). As a note, his son, T.J. Miller, is well known across the world as a movie actor and comedian.

NS: "In conventional terms you show all the signs of success, but on your own terms, what is your own definition of success? What do you see as your great achievements?"

KM: "Well, I've been successful in my career. But I've got four kids, and they all got through college. All of my grandkids are here in town, and my two youngest kids are starting their own careers. I'm happily married, and I've always had real identity outside law practice. That's success as far as I'm concerned, I've got a good family life."

NS: "I know that you won't take just any case regardless of how much money there is in it. How do you determine what cases you'll take and which one's you won't get involved in?"

KM: "I can't really get into a case unless I believe in it and can express an opinion that will be helpful. And if I can't do that, I tell the lawyers as early as I can and I tell them why. I'll take a case if I'm comfortable in defending the position a lawyer is advocating. I've worked on both sides, on behalf of a company but also on behalf of an individual, depending on what I think is right."

NS: "You like your job, you have this freedom and you're not carrying a lot of baggage around-"

KM: "Over the years I've set it up so I can work at home when I'm not in court. I don't have a receptionist, I don't have a secretary or a staff. I was once the managing partner in a twenty-five attorney firm. I was just bogged down. Now I can just concentrate on the job at hand."

NS: "So how did you arrive at this place instead of having this big unwieldy firm under you like a giant pyramid?"

KM: "I just decided at one point that I was going to work at home and watch my kids grow up. I dropped out of the boutique law firm business with all the staff and all those headaches. It took a little time to make the transition, but I moved into my home office, and that was seventeen years ago."

NS: "So a big part of your success is how you feel, so you enjoy what you're doing. It's just not a pile of gold in the vault."

KM: "Almost every day I play handball or squash in the middle of the day. I might want to have a beer with the guys in the locker room and shoot the breeze. I couldn't have that kind of lifestyle if I was still in a big law firm. My daily commute is down from my garage down to the athletic club. My social life isn't lawyers. That kind of schedule is very important to me."

NS: "Let's talk about cars just for a minute. You're a big automobile enthusiast."

KM: "I grew up in a little town in southeast Kansas where all the 'cool' guys had '57 Fords and Chevy 409's. My parents didn't care anything about cars so I had to figure out a way to make a 1956 Buick station wagon cool, which is impossible to do. I would take the hubcaps off, and put them back on… (laughs) So these days, me and a lot of my friends are in the car clubs. I've been a driving car instructor for the Porsche club for years and I've instructed at Ferrari national events."

NS: "Okay, back to brain- Part of my work revolves around educating people about how they can consciously overcome negative emotions in a threatening situation, and turn off the limitations of the fight or flight response, and replace it with intelligent thinking that is better for survival and positive emotions."

KM: "So that's what I do on the witness stand when I'm being cross-examined. The people who hire me give me very high reviews for being able to handle quite hostile questions with a professional response, as opposed to either the fight part, which would be to get combative, or the flight part, which would be to say, 'Oh I don't know, I guess that you're right…', to kind of give up.

NS: "So your opponents are trying to get you to click your amygdala backward on the stand?"

KM: "Yeah. Either they want to have me give up or they want me to come off as hostile and combative and unprofessional. So if they can get me to do either of those things they'll score points with the jury- and I lose.

What they don't want is for me to be calm and explain things in a rational and sensible manner. But that's exactly what I'm good at."

NS: "So your ability to consciously override that lower instinct for fight or flight is why you are so good at your job. You would be a failure if all you did was to react to the pressures on the stand. "

KM: "Right. My whole attitude is not to be aggressive or combative, and not to be emotionless, but to bring logic into it."

NS: "You've just described what it means to circuit forward from the primitive reactions of the reactive reptile brain and to move forward into the frontal lobes. Your logic and cooperation is what is going to keep you afloat, where as your combat is going to sink your ship."

KM: "Yep. If you get baited into that combat when you're with another lawyer... What lawyers like to do is to phrase everything in the negative, and I flip it over and answer everything in the positive with a "Yes", still saying what I want to say, and they never catch on! (laughs)

I just decide that I'm going to consistently take the high road. But I do that in my private life too. It's a much better way to live in every respect."

AMYGDALA TICKLE #34- "Time Out"
Make time to get out of the Rat Race- for a time every day. If you are a prisoner of your own making, what's the point of living?

Make sure and take a Time Out. Spend it with your family or a friend. Take an hour to get some exercise and fresh air. Relax. Depressurize before you blow up. Don't be lashed to the oars – Captain your own cruise ship.

Mark Foster
Professional Percussionist, Music Faculty Metropolitan State College

Mark Foster can be heard regularly on stage and in concert productions in the Rocky Mountain region. This interview began in his basement, which is a virtual musical laboratory and warehouse of countless percussion instruments of every type, many of which are unique and which he has constructed himself.

MF: "This is called a water phone... I got this specifically for the Broadway Musical "Wicked". You put water inside this thing and use a violin bow to play it..."

NS: "It looks like a UFO!"

MF: "There's just one guy that makes these, named Richard Waters, in fact." [Mark plays the instrument to a very haunting howling effect.]

NS: "Oh man! I can see why you would use that in the Wizard of Oz for the witch- that's really incredible."

NS: "So anyway, I'm always aware of how people perceive 'who's a good musician'. Very often people mix up name recognition and fame with what truly makes an outstanding talent.

The really outstanding musicians that I know in this town also value other things in life. You, for example, are a very avid gardener. You've found this healthy balance where you can still pursue your art, but you've maintain your home life and you've recognized the value and importance of it. I mean, you're going to live to a ripe old age."

MF: "In my estimation, the guys I know who are really successful are down to earth people. And also, the guys I know who are free lance musicians are a generally a very happy lot.

That's the thing I like about my life- I can go and play in a symphony, and then I get to play show music, and then I go play ballet- I just finished playing tympani in Swan Lake, that was really cool- then I play drum set kit and that really grounds me.

I noticed that when I'm sitting down at the drum set, I don't forget anything no matter how long I'm away from it."

NS: "Like riding a bicycle?"

MF: "Yeah. So I started examining it and asked myself, 'How did I learn the set as opposed to the other instruments?' And this light bulb went off-

What I realized was that when you learn drum set you don't start off with anything too complicated, just the basic beat. And the first rule is to *get that thing to feel really good.* You can play along with a record, and your thing is to get your time to feel as good as the other guy's.

It's not about complexity; it's about simplicity- and trusting that it can feel easy.

I started to compare that to my training with mallet instruments. The lesson I learned from mallets was virtually the opposite approach- I remember getting these mallet pieces was that you find the hardest part of the piece and you go to war with it- and you defeat it!"

NS: "That's very reptilian!" (laughs)

MF: "But what I learned about playing mallets was that 'This was always going to be *hard*' at a very deep level. I was never able to trust that I could do it because I was always *fighting it.* I never learned to feel good with it.

What I eventually learned for myself was to look at it and find first what looks easy and start there. And explore and always work from the easy familiar parts and go from there to the parts that are different.

NS: "You're comfortable from the beginning…"

MF: "Right- you're letting yourself explore it in different ways instead of going through it from page one and going through it all the way across to the end. You find something here, and then you look for something that's like that- You're letting your mind explore in that it could have been put together in all these different ways, because you're putting it together in different ways as you're learning it. You might practice it backwards.

In my teaching I've learned to have my students take what's on the page and changing it to be different and something simpler. Then your brain remembers it, and all you have to do is change this one little thing to make it like it is on the page. The trick is in seeing the simpler things that are hidden by the complexity. You look for the patterns and the simplicity of things underneath everything."

NS: "You look for the simple things."

MF: "You simplify and simplify all the time."

NS: "So what is the effect of this on your students?"

MF: "I see it in what my students can do and the age at which they can do things- My high school students play the kind of things I played in my senior recital in college. They far surpass me in some of the things they're doing. I've got this one kid now in the Manhattan School of Music, he's just an amazing mallet player and I'm never got match that kid!

In your life you're trying to uncover those things that still bind you, the things that don't allow you to be free. My philosophy these days is all about release, whether it's psychological, or physical, or emotional, and how all these things are connected. What's the motion that allows the next motion to be free? What's blocking the freedom?

It's all about getting loose, being able to relax, and having freedom of motion. Then it's so much more efficient- and enjoyable."

AMYGDALA TICKLE #35- "Dendrite Drum Delight"

Many percussion instruments are easy and fun to play and will instantly get you flowing into frontal lobes rhythm when you otherwise might be stuck in a pit of uncreative combative reptile brain mud.

Get some bongos, a hand drum, or other percussion instrument. Whip out your favorite record and just bang along. No lessons needed. Connect with a drum circle and run rings around the un-musical decapitated troglodytes in the cave next door.

Paul Conly
Rock Musician, Composer, Electronic Music

Paul Conly is recognized for his pioneering music and his band "Lothar and The Hand People", the band name coming from their use of an electronic instrument they named "Lothar". The band was notable for being the first rock band to tour and record using synthesizers along with regular instruments (1965), and for their part for inspiring other electronic music-makers who could soon be heard across the radio airwaves.

NS: "Tell me about Lothar and The Hand People."

PC: "This band was started in Denver, and then we moved to New York City where we had a contract with Capitol Records."

NS: "The Beatle's label."

PC: "They released our records nationally, and we had some bootlegs in Europe. We made the top ten in some locations. What got me interested in joining the band was their Theremin. It looks like a box with two antennas on it. Our Theremin was made by Robert Moog of Moog Synthesizers."

NS: "Can you give me a couple examples of famous uses of a Theremin?"

PC: "Movie soundtracks for "The Day The Earth Stood Still", Alfred Hitchcock's "Spellbound". People always think the Beach Boy's "Good Vibrations" used a Theremin, but it was a bit different than ours. On a real Theremin you don't actually touch it. The proximity of your hands to the antenna actually changes the notes and the volume."

NS: "The example that comes to my mind is Jimmy Page and 'Whole Lotta Love'. But your use of the Theremin actually pre-dated his use by about five years. But what about your background- how did you get into music?"

PC: "I grew up in a musical family. An early memory I have is volunteering to stand in front of my second grade class to sing "Twinkle Twinkle Little Star." I was always rewarded for being musical."

NS: "When I was five years old my uncle paid me a nickel to sing "He's Got The Whole World In His Hands". So I can blame my uncle for my first paid gig as a musician, which set me down that long dark path."

PC: (laughs) "You're very impressionable at that age. One turning point for me was in high school. I had an English teacher who had me write essays, and all I got was C's, not good grades. Then there was an all-school talent show for which I played guitar and sang. It turns out, that teacher was the faculty person in charge of the show. After I auditioned and played the show, every grade I got from her from that point on was an A+. (laughs) I don't think it was just some coincidence."

NS: "I'm certain that we are always guided in life from both the external rewards we are given, but also as the internal rewards that we get inside our own heads for doing something. You realize, 'Oh, that feels good- I could do that again', and you keep developing that and go down the path that provides the reward.

But what about the obstacles, disappointments, and frustrations that come along on any artistic path? Being an artist of any kind isn't just a bed of roses. How did you keep going through all the inevitable challenges of playing music?"

PC: "There was always that *self*-reward in it for me. Once you establish that internal self-tickling then the patterns are established from that. But in my family, there was a lot of encouragement for that, we always got together and played for each other."

NS: "Like the Bach family-"

PC: "Haha- yeah. But certainly not like performing in front of strangers or your high school. But the first time I played with Lothar I'll never forget it. Up until that time, people really didn't pay that much attention to me.

On that first gig with Lothar however, there were literally hundreds of people all standing and staring at me, standing just feet away. It was disturbing!"

NS: "I remember reading of Jimmy Page [Led Zeppelin] and his stage fright which preceded every performance throughout his life. But he was always able to get over that momentary difficulty knowing that it would pass. That's a function of the frontal lobes; the ability to overcome a present difficulty and move down the time line past a temporary rock in your shoe."

PC: "I remember at that talent show, I had terrible stage fright. I had butterflies in my stomach and my voice had this drastic involuntary vibrato-[makes warbling sound] But after I was into for a minute, I got over it.

When I was in Lothar, that band could be the best band in the world- or the worst. The funny thing was, we never got heckled when we were awful-only when we were playing good. I think it's because heckling is really a way for some guy out in the audience to get attention back from his girlfriend who is maybe paying too much attention to the guys on stage." (laughs)

NS: "Do you remember in The Beatles' movie *A Hard Day's Night* when the band would get chased down the street by hoards of wild girl fans? You had that happen to you as well, didn't you?"

PC: "It was a phenomenon of the time, and the age of the audience. And to tell you the truth, when it happened it was pretty scary. I mean, there were times when the girls came at you and they had scissors because they were hoping to cut your hair and take it home as a souvenir. But there's a difference between one or two of them and then twenty of them pointing with scissors. There's a reason for running! You start realizing that they're armed and dangerous!"

NS: (laughs) "Society tends to see that kind of thing and think, "Oh wow, wouldn't that be great to be one of those guys?" But that's not necessarily what its like."

PC: "I'll tell you another story- we played at what should have been an ideal situation, at the Fashion Institute of Technology. For us, this meant playing on a stage of about eight-hundred totally beautiful young women in the audience... If you're in it to meet young chicks, this is the job for you! It looked like the ideal situation.

So we were on the stage, and played well, and got a good reception. And they asked us to stay and meet the girls and sign autographs, which we did. But as it turned out, I really didn't enjoy it.

What they perceived of us was really not who we were. It had little to do with what I thought of myself as a musician. It was empty adulation.

Our band wasn't what I would call really famous, but we were close enough to see what "famous" was all about. I just didn't care for that at all. It didn't mean anything.

What music expresses is something very profound, something that perhaps we can't even put into words. To have that reduced to something like empty hero worship- it's not for me.

I once had this girl say to me, 'You're one of the Hand People- and I'm just *nobody*.' The idea that because I'm playing music- that makes me a better person- I don't accept that.

Everyone is valuable to me. I started to resent any idea that puts people on pedestals. Maybe that's the reason that our band didn't become that overwhelmingly famous- there was a part of us that felt like it was an ego trap. Because you can get trapped thinking that you're exceptional, and that you're a prima donna of some sort. People like that don't interest me."

NS: "When did the band thing all end?"

PC: "The group ended with the '60s, in January of 1970.

I was friends and hung out with Sam Shepard [actor/playwright Pulitzer Prize, 1979], who asked me to do incidental music for a couple of his off-Broadway plays. I was sponsored by ARP synthesizers then, and I introduced to the world in what was probably the first stage play that featured synthesizer music.

After that I moved to Boston where I managed to get access to the Harvard University computer lab, quite an exceptional thing in those days. Their "mini-computer" back then was the size of a large refrigerator, and we used it to create computer music. I had a $100,000 ARP synthesizer prototype back then, which was quite an extraordinary thing. There were only maybe two or three other places in the world that could do something along those lines at that time.

At the same time, I had all of these incredible opportunities held in front of me: to record with my ex-band mate Tom Fly [Woodstock album] in New York; ARP was offering to set me up in an eight track studio of my own in New York and to have me tour college campuses showcasing their instruments. But I was really more interested in experimenting and doing my own thing. I had a million dollar facility at my disposal in Boston, so I turned everything else down."

NS: "So you had this mad scientist period for how long?"

PC: "About a year and a half. But the cracks started to show when we had this accident and fried that $100,000 ARP synthesizer one day. We were creating this hybrid system, where you could create sounds on the computer and control it with the synthesizer. So this synthesizer was in one room, and the computer was down the hall in the psychology department where I used to ride the elevator with B.F. Skinner.

One day the ARP guy is there and he accidentally plugged the synthesizer into the wall socket by mistake and he sent 120 volts down the chassis of the synthesizer. There was one of those things like in the movies where all the lights dimmed in the place, and the place was filled with this horrible smell. This would have electrocuted anybody if they had been touching it. That was the beginning of the end of me and ARP!"

NS: "You've written a lot of music for films and commercials..."

PC: "Yeah, I've also done music for over two-hundred films and videos, I stopped counting. I've done some things for Sesame Street that are still being shown."

NS: "As you look back, you've had a lot of opportunities, and you've said yes to some things and no to others. What was the principal key point in deciding what to do?"

PC: "It was always the desire to experiment and find something new. I've always just been obsessed with creativity and what would happen if, for example, you hook up a synthesizer to a computer."

NS: "Well, that's something you see everywhere now. You go into Guitar Center and that's what all the keyboards are about, a computer married to synthesizer tone generators. You were doing this almost forty-five years ago. But you were really never attracted to fame or wealth, and you didn't achieve the kind of notoriety that we typically associate with success. But you were still successful on your own terms, because you achieved your goal of mad musical scientist."

PC: "That's a great way of looking at it. Oh by the way, check this out- (Paul gets up and shows me a Zippy comic strip framed on the wall). Bill Griffith signed this for me in 2011, it's called 'The Day The Theremin Stood Still'."

NS: "Oh yeah, here in this last panel it mentions 'Lothar and The Hand People'."

PC: "Back to what we were talking about- not being too famous is good because you get the status of forever being in a cult group. (laughs) But I'm proud of what we did. People say that The Beach Boys were the first ones to use a theremin, but, uh, no, we were, and we were actually performing with it live."

NS: "I'm fairly well versed with the history of the Beatles, arguably the most famous band ever. Although John Lennon was proud of the band in a certain sense, he was always aware of the fact that at some level he felt that they had sold-out artistically. He had often expressed great dissatisfaction with their recordings. George Harrison once said, 'The worse thing in the world is being a Beatle', referring to the huge personal price he paid for all of that fame. And of course, we all know the price Lennon paid for fame.

So our perception of success can be very different from the person who is in the middle of it."

PC: "Yeah, I know that I never really wanted it that much, and the taste of it that I did have made me feel even more strongly so.

Our first producer really tried to make us commercial, and I really felt he was wrong in two main ways- By trying to make us commercial it was like trying to fit a square peg into a round hole. It wasn't going to work artistically.

But the other side of it was that by grinding off the corners he destroyed the essence of what made us appealing to people. What appealed to people when we played was that experimental silliness and the fact that we were having fun enjoying it, and that we weren't even trying to be commercial."

NS: "Humans do that- 'Oh that works! Let's copy that!' The copy-cats are never as successful as the originals, at least artistically."

PC: "Honestly, I've thought many times that if we had been more successful then I probably would have not survived in the same way that Kurt Cobain did not survive. We just barely got out of it alive anyway. I'm serious about that. If we had made it into the major leagues, I probably wouldn't be here to talk about it. If you can survive, house and feed yourself however modest, based on the success of your work- that in itself is a definition of artistic success to me.

I also think that if you just if you persist in what you're doing- that's an important measure of success. Popular acclaim? That's another thing entirely. I think the price to pay for that is too much for a lot of people. Certainly for me." (smiles broadly)

AMYGDALA TICKLE #36- "Off The Boring Beaten Path"
Instead of listening to the Same Old Thing- as much as you like it- ask around and EXPLORE with your frontal lobes and find something NEW.

Although it might take some trial and effort, your brain gives you a bigger reward when you explore and hit gold, something that just won't happen on replay.

Frank Zappa
Musician, Composer

Speaking of "Off the beaten path", contrary to what most people might think, famed composer, guitarist, and iconoclastic rock star Frank Zappa lived more or less a middle class life style in the suburbs of Los Angeles.

I met and spoke to Frank Zappa when I was just starting in my own music career in 1972. He was polite, was willing to answer any question, and gave me the names of several books on music theory and composition. His parting words reflected a common sentiment, that creativity is generated and perceived through the Big Magic I, from a source outside one's little self:

"Remember you are employed and working for the Muse."

AMYGDALA TICKLE #37- "Who Are The Brain Police?"
Frank's music covers many genres- rock, comedy, avant-garde, classical, jazz, pop, country- you name it.

Access your favorite online search engine, and simply type in the kind of music you prefer. Chances are dollars to decibels Frank wrote one or many world class examples in that style, and did a great job of it to boot. Give it a listen to tickle your tympanic membranes as well as your amygdala.

Jean Massey
Public School LTA, Public School Guidance and Behavior Counselor

NS: "I wanted to know how you apply what you know about how your brain works to your life. Did you say that you live in the country in New York State?

JM: "I was born and raised in New York City, and in my mid-thirties I moved out to have a family. So, I'm now living in a very remote section of the Catskill Mountains, in a little community with about 300 people, and I've been here for about twenty-six years. All my life I had wanted to live in this kind of environment, and I got lucky."

NS: "How does that environment make a difference in your mind-set?"

JM: "It wasn't so much the environment, but rather a couple of changes in my life and a traumatic experience- My husband left me some time ago and I had two young children to raise. It was around that time that I started looking into things differently than I had ever done before. I finally paused and asked myself, 'What am I doing? What is the purpose in all of this? Why did I go in this direction? Why did all of these crummy things happen to me?'

At that point I started looking at a lot of things, I started reading a lot and looking into different programs- there's no TV up here."

NS: "That's different right there!"

JM: "Once my husband split, I had to find a job. First I had a job in a little café, then in a gas station store, then I worked for the Girl Scouts organization. I was in my mid-forties and knew I needed to bring in more income. I volunteered in the schools here for a while, and they eventually offered me a job as an LTA. I am the school disciplinarian, and I've been doing that for fifteen years."

NS: "Ah, so when kids are bad, they send them to you."

JM: "Isn't that wonderful? (laughs) I love it.

I was a wild single for most of my life. I had a traumatic life- I lost both my parents, my father when I was three and my mother when I was seventeen. So I was a survival expert you might say, but in a reptile way."

NS: "Tell me a little bit about that pre-country life."

JM: "I struggled, but I didn't see it as struggle of course. When my friends were playing with Barbie dolls, I was climbing trees. So nature was always a very important part of my life. But as I grew into my teenage years and started hanging out on the street corners of Brooklyn, that had to disappear because I wouldn't have been accepted. But something must have survived because I ended up here.

I ended up in a place- my house now- which is in that interface between forest and field, which was very important to me as a child. Although I didn't write it down, I visualized that as a kid, and I have that now.

After my mother died I looked at life differently. I decided that street life wasn't what I wanted. So, I joined an environment group called "The Hudson River Sloop Restoration", which was led by the famous folk singer Pete Seeger. I worked on a boat, and eventually lived on that boat and cooked for a season. That changed my life. It turned me to folk music and led me back to the country and to the river and to nature, things that I had turned away from during my crazy teen years and threw out the window."

NS: "How did you discover the amygdala and get into all this brain stuff?"

JM: "When I was hitting the bottom, after my husband split, I remember being in the car, my hands tightly gripping the steering wheel driving down this beautiful road on this beautiful mountain. Everything was *perfect* around me- but my life *sucked.*

I didn't have money, I was fearful, I was alone with two young kids, and I was waking up in the middle of the night, sweating and freaking out. I used to call it the Hebee Jebes. I was freaking out in the middle of the night, every night.

I remember one time that it hit me- I'm clutching the steering wheel saying to myself, "This is against any law of probability, that all of this would happen to me. There's got to be some reason behind this.

I went into work and cried my eyes out. I was saying to myself, 'I can't stand this, I don't want to do this any more...' But I couldn't just end my life because of my kids, but I was down at the bottom.

And all of a sudden I hear this voice scream out at me that says 'Stop it!'

I looked up to see who said it, if it was a kid or somebody else, but there was nobody around. I went back to my crying after a couple of moments, because I was a cry baby back then, and I was crying my eyes out- and there it was again, the same voice came back again and said, '*Stop it!*' That got me thinking, 'What the heck is going on here?'

I started looking to find someone who could provide me with some help and counseling, but in this remote area there wasn't anything. Eventually I started looking on the Internet, and there's a million self-help sites out there- I started reading. You need a lot of other ideas, and see what other people are doing and thinking, and then you can develop your own ideas of what's going on.

So I tried different things and I was getting better. So I kept looking and reading. But I also believe in the idea of 'I don't know.' Because as so as you believe that you do know, then you get stuck, and that's when I'm blinded again, and I've got that wall in front of me again."

NS: "I was just listening to Robert Anton Wilson, and he was saying something right along those lines. As soon as you decide you believe in something, you stop learning..."

JM: "That's right. Anyway, I was doing all these things, and every day I would come home, and go to this special spot outside near my house, and I would look up in the clouds, and like magic, there would be this clearing of the clouds and there was this big patch of blue sky above me, and it would happen every day. I was doing this every day when I came home."

NS: "So you saw this cloud thing, and you weren't consciously doing it, but you noticed that it was happening."

JM: "Yes, it was just happening. So I get on the Internet, and found a page about 'cloudbusting', and then started to read about the amygdala, and I printed out the articles about it.

I had a friend come in, and I tried to explain it and told her, 'I do it and I just feel so comfortable, its just a happy feeling and I'm so satisfied- it's like I've reaching this pinnacle in my life- I don't know what's going to happen tomorrow, but I'm just not worried, I'm not scared!'

So my friend went home, and she was able to do it, and she showed her husband, and she was just beaming love and happiness. Then, I finally landed on your web site, and you went into this great depth explaining things about tickling the amygdala, and that's where all this stuff was coming from."

NS: "You mean about cloudbusting, or tickling your amygdala?"

JM: "About tickling your amygdala, and the brain. That's when I just started consuming all those books, because I felt that everything was coming from the mind and the brain, that's where it all centered around."

NS: "So do you routinely tickle your amygdala?"

JM: "Oh yes, I still do."

NS: "Just explain how you do it, and in what circumstances, how often, and what happens when you do it."

JM: "I go outside, I have a chair and a spot, and I do it almost every day outside in the summer. I set a specific time."

NS: "So you don't do it during your daily activities?"

JM: "Oh, I do it during the day too, if I think of it, if it comes to my mind."

NS: "How do you do it?"

JM: "I imagine a feather- like I might be tickled on a spot in my body, but I visualize it happening inside my brain in this spot on the amygdala. I've tried other things, but that does it the best for me. I do them both at once."

NS: "You can do this right now while we're speaking?"

JM: "Oh yeah, that's easy."

NS: "That's great. A lot of people don't understand that tickling your amygdala doesn't necessarily mean that you're getting instant nirvana, but it means basically that you're changing the direction of energy in your brain- so that instead of energy being stuck in your reactive reptile brain, you're moving the energy more into your frontal lobes, merely by using your imagination in this positive, pleasurable way.

And your frontal lobes allows you to see around the corner, and avoid a reality hangover later on which your reactive reptile brain can't do, just seeking momentary pleasure for its own sake.

The point is, everybody does it their own way. That's why I'm interviewing so many people for this project. Different people find different ways of tickling their amygdala.

JM: "I understand what you're saying about the frontal lobes tickling the amygdala in a lot of different ways, and that's wonderful. I didn't really get that before."

NS: "Oh sure- I was talking to Stevie Wonder's piano tech last week, and he was telling me the same thing- he had a tuning fork and he would hold this tuning fork up to his forehead or temple, and tickled his amygdala that way. Some people do it one way, others do it their own way."

JM: "Oh wow... so it's all frontal lobe, that's great."

NS: "In dealing with your disciplinary job in the school- how has tickling your amygdala changed things?

JM: "I never tortured these children."

NS: (laughs)

JM: "These children are just rebelling. The kids who come to me are rebelling against brainwashing. I always thought that 'The Terrible Twos" is just the first sign that kids are rebelling against this terrible way of thinking that we live in this world, the 'War Mentality', the reptile mind.

The kids who come into my room are children that are *chosen*, they're fighting this- I always call them 'The Best Kids In The School'...

NS: "Whoa! Hold on a second there- you just said something so unique and so profound..."

JM: "Okay..."

NS: "So the kids are being sent to you for discipline, but you don't see them as being the *bad* kids..."

JM: "Nooo!..."

NS: "You see them as being the *best* kids in the school! What a revolutionary concept!"

JM: "It's the *society*! They're the only ones that are standing up and saying- they're *refusing* to follow this old regiment that we're in now. They refuse!"

NS: "That is such a brilliant and perceptive and creative attitude!"

JM: "But it's true, though- aren't they? The kids who all just sit in their desks have already been conditioned years ago... You can't blame anybody, because we've all been through this, we've all been conditioned. Both my kids are honor students, so don't get me wrong, I'm not putting down anybody.

But the kids sitting there, getting the perfect grades, getting the awards- everybody is applauding them, they're getting assemblies to applaud them, they're in the newspapers, they get the medals, they get extra field trips- and *it's the kids that come to me that are the movers and the shakers*. They're the people that we have our hope in- *if we don't kill them*."

NS: "I'm speechless- That is such a profound and far seeing perception. You are light years ahead. You are light years ahead..."

JM: "I've been working very hard at this, and that's why I'm in that happy place most of my day. I don't do this for the money- although a little more money wouldn't hurt! (laughs) I don't make much money, and I don't have much. But I have everything else. I have everything I want in my life.

You have to recognize what's out there, you have to see the truth.

I have people at school who wonder, 'What do you do with these kids?!' because I have less kids who come to me every year, less and less. I believe you project your life, and I'm projected something better and better.

I have to tell you something, both my kids went off to college- my daughter got into a very prestigious music conservatory, and hated it. She came back and is going to a local music school here instead and loves it. My son quit college in his third year, he was going to become a teacher, and all my friends went, 'Oh my god!' and told me how awful they thought that was- and

I'm grinning inside, because he knew what he wanted, and he had enough of it and he knew what he didn't want to do. He was listening to himself."

NS: "This is not to say that everyone who does well academically, or fits into the standard scholastic mold is wrong either- some people thrive in that way."

JM: "Yes."

NS: "The mistake we make is when we see someone who does not fit into the mold, we label them as a failure. Steve Jobs was a dropout. Albert Einstein did poorly in his schools. We have to recognize that everybody is different. An A+ student doesn't necessarily mean that you are an A+ human being."

JM: "But it is the way society thinks it is."

NS: "This reminds me of something I ran across recently, someone said this, 'I don't think there are any bad people out there, there are just bad ideas.'"

JM: "I tell my kids that it's not *them*- and it's not their teachers either."

NS: "You just need to think about what you are doing and where you are going."

JM: "Yes. We are all just starting to wake up."

AMYGDALA TICKLE #38- "Each One- Teach One"

Join the Brain Revolution. Get an apple and a couple of feathers. Hand them out to some kid you know- or that big kid occupying the office chair next to yours at work.

Explain how you just learned that the human brain is made up like an apple, with the big juicy frontal lobes right up front-

AND that anybody can tickle their amygdala Goodie-Spot with a frontal lobes feather to find the right direction to go.

Sky Wise
Teacher, Artist, Musician, Author

NS: "I met you about thirty-six years ago in music school, where we were both attending. But you already had another degree then- in fact, didn't you have two prior degrees?"

SW: "Yes. I had a degree in psychology from the University of Kansas, and then one in art from the University of Colorado. Yeah. I've studied some different stuff. I've just been all over the place, just like Amy Tan's mother said to her, 'You're like water, spread all over...' Bit it's just part of my personality. I find that when I do painting, that my music background gets into it. Or sometimes I think about literary things that inspire me visually."

NS: "Wasn't Thomas Edison a jack of all trades? Nobody criticizes him for that…"

SW: "But he was a genius and persistent. The willingness to sleep on a hard surface all night just to stay in his studio and keep with and idea until he solved the problem… So I think that any artistic and inventive process requires that you stay with it long enough so you get some closure. And you have to keep open to influences."

NS: "Well you certainly stuck with each of those long enough to get your degree in each."

SW: "Yeah, but eventually I discovered that in the public school system I was a fish out of water. I'm a free thinker. I want to do things my way."

NS: "Well you weren't just a teacher- you're an artist and a musician…"

SW: "Yeah, and that put me on 'The List'- *and* I agreed to become the union rep, so that was my fatal mistake!" (laughs)

NS: "But you taught in the public schools for quite a long time."

SW: "Fifteen years. And I did have students that came back to see me years later. One was a young Samoan islander, and he became a star of sorts. He told me, 'Thank you Mr. Wise for encouraging me in music.' So I was glad for any of my kids who went on and did things with their music. But I can see that now that I'm better off producing my own stuff."

NS: "So tell me about your current project, your book."

SW: "This came out of my school experience. Kids were trying to learn language, but the school dictionaries were atrocious, and cheap and torn up. The problem is that often kids can't find the word they need in the dictionary.

I saw so many kids who could read, and could read fast- but it was just 'Word Salad'. They didn't understand what they were reading! If they don't understand 80% of the words they are reading they're not comprehending anything.

So my dictionary is different in a lot of ways: It's illustrated, and it presents verbs in context and in past tense so kids can actually find the word they're looking for.

Part of the creativity is finding examples of English words that are in common use, but there are unusual or striking examples. For example; the literal 'He stole a book' versus the uncommon 'She stole a glance'. It's the second example that people are less aware of, but they're interesting usage of the language. So I want to include more and more of those in my book.

As far as illustrations, I became more interested in facial expression, and to include more faces and emotional expressions in the book, to make it really unique."

NS: "You saw a problem in the educational system- Why didn't anybody else see it?"

SW: "I think they did notice it, but they couldn't find the proper solution for it because it wasn't in the standard curriculum. And I'm also more sensitive to meaning in language because I'm sensitive to the sound of things because of my musical training. People just aren't sensitive enough.

One day we were grading kid's papers, and my supervisor is telling me 'A lot" as in 'a lot of things' is one word.

NS: "He's telling you 'a lot' is one word?"

SW: "So instead of being diplomatic, I just said to him, 'Let's look it up,' and I proved him wrong in front of a lot of people. I learned that you don't prove your boss wrong in front of a lot of other people. (laughing) A year later he was trying to get me fired." (laughs)

NS: "So you know all about the reptile brain and the ability or inability for people to tickle on their frontal lobes in the work place..."

SW: "I have to say, that all you need is to have three people to poison a whole environment. I saw extremely intelligent people act in negative and uncooperative ways that were bad for the teachers and bad for the students.

It's bad if teaching is just like workers on a conveyor belt on an assembly line. It's bad when no one is thinking about anything very deeply. We're reliving the environment of Dickens's *Hard Times*. In his viewpoint, they were trying to kill imagination back in the 1900's with industrialization. Kids had numbers instead of names.

When you have emphasis on data collection in the schools it leads to cheating. It leads to infighting and fractionalization.

It was hard for me to leave the schools, but in its own way, it's proven to be a motivating factor for me to finish my book."

NS: "Necessity is the mother of invention."

SW: "I believe that is very true."

AMYGDALA TICKLE #39- "The Expanding Wordiverse "
Use a new word to 'edutain' your Left Brain pre-frontal cortex.

Learn a new synonym using a thesaurus to expand your lexicon and expand your brain and amplify your verbal amygdala tickles.

Dr. Stanley Kerstein, M.D.
Physician

Dr. Stanley Kerstein is a general practitioner with a large and devoted patient following. He's admired for his broad knowledge, exceptional down-to-earth style, and the warm reception and ease that he imparts at every office visit. He might be equated with the old country doctor, but in a modern urban setting.

NS: "For you as a physician, what is the key for being effective, and a successful in helping people heal?"

SK: "Listening. Listen to what your patient is telling you. Also, to give patients to have insight as to what is ailing them so they can help themselves. Communicate to patients why you are doing things.

One of the criticisms of our profession- in fact it happened just today- when someone was referred to a specialist, they came in and he didn't explain anything to my patient who then had a very unwholesome experience. So, communicating and listening to your patient is of key importance."

NS: "I knew that about you from the first time we met, you were interested in me as a person, not just a potential carrier of the latest most popular disease."

SK: (laughs) "If it were up to me, I'd spend an hour with every patient because I like talking and I like hearing about people's lives. I learn from other people, even from writers like you on occasion. (laughs)

But seriously, the interaction with the patient, that to me is the most important thing. Also, allowing the patient to participate in the decisions about what we're going to do. In most situations there's more than one way to do something."

NS: "So you're talking about a cooperative spirit?"

SK: "Yes, like, 'We're going to do this together, there's more than one way to do it.' It's not one-size-fits-all.

There are people who want to do things a little differently, and to that extent that I think its okay I'll do that. But if I see you're making a potentially life threatening decision or error, I'll certainly speak up strongly."

NS: "I've been talking to all kinds of people and trying to find that thread that guides people to discover what they find is the right path for them- even if they may not consciously know what that process is inside their head."

SK: "I was probably attracted to medicine because of the science, and the desire to help people, and it sounded like a fun gig. I've always been attracted to science, and I've always been interested in people and doing things with people.

But radiologists are probably interested in the technology, surgeons are probably more interested in your inflamed gall bladder and couldn't care less about you as a person. (laughs) I mean, they're nice people but they're probably more interested in getting that thing out of you then sitting there kibitzing with you about your life.

But I like hearing about what you do, so I think that's fun. People do things, you learn about the world. I see a whole spectrum of people who come into the office- it's a kick."

NS: "Well, that's the key right there- you're tuning into what the reward is for you, the positive feedback that you're getting inside your brain."

SK: "To some extent we all have to do some things we're not crazy about, and that's the reality of making a living. But hopefully along the way you're doing something that you really enjoy.

But there are people who work and do things because they ended up there and they have to make a living, but they hate what they do."

NS: "It seems to me that those kinds of people are never as successful as those who actually enjoy their job. How long have you been practicing medicine?"

SK: "Man- I've been practicing for about thirty-six years, a heck of a long time, I'm sixty-six now." [Doesn't look it!]

NS: "What? That's incredible."

SK: "I enjoy what I do. And genetics helps." (smiles)

Dr. Cheryl Chessick, MD.

Psychiatrist and Psychiatric Clinician, Director Depression Center, University of Colorado Medical Hospital

Dr. Chessick continues an outstanding career in the field of psychiatry and psychiatric clinical research that would take several pages of this book alone to cite, and for which she has receive numerous honors and national awards. Among these, the Irma Bland Award for Teaching, American Psychiatric Association (2009), Far Beyond the Ordinary World Class Care Award (2006), the Jay Scully, M.D. Award for Distinguished Teacher (2004, 2009), Distinguished Fellow, American Psychiatric Association (2000)

From 2008-2009 she was the Chief Medical Officer, University of Colorado Hospital Department of Psychiatry, and is currently the Director, Women's Treatment and Studies, Depression Center, University of Colorado Medical Center.

NS: "Can you briefly describe your current work?"

CC: "I'm a psychiatrist here at the department, and at the Depression Center part of our goal is to find better ways to treat mood disorders for the State of Colorado. We look at preventive measures, what are the best practices, and how to get access for people."

NS: "What sets someone who comes into the clinic apart from a person who feels like they can go on with their daily life without too much interference?"

CC: "All of us have our moods, but we tend to carry on with our daily activity. But people who come in here are having symptoms to the degree that it interferes with them carrying on in a normal way with these things as they were before."

NS: "Is it possible to make an overall general statement about people who suffer from that debilitating kind of mood disorder?"

CC: "There are some general things, but everybody is an individual in how they deal with themselves and their environment. It may look a little different for each person, but when people feel poorly they tend to withdraw from their own normal patterns and the way they interact with people. And the more you withdraw, the more you withdraw, and it goes round and round and round."

NS: "You pull back into your own shell…"

CC: "Right."

NS: "I would expect that when you do diagnoses, there is sometimes a chemical and physical imbalance that you treat or begin to treat with medicines- but at other times you see it as a behavioral process that can be changed with attitude. Are these separate? Or how can you make that determination?"

CC: "It's a complex question. There are limitations that we have in mental health that are unique. We don't have an ability like you might have in testing for a glucose level, or as you might test for diabetes, or test for a thyroid condition.

We're at this point where we know that it's just not a chemical, but that it's also about the circuitry in the brain that may or may not be working correctly. When you work with someone, it's probably a mixture of both."

NS: "Is it fair to say 'how you think' in fact changes your brain and your brain chemistry?"

CC: "It can."

NS: "Is it fair to say you can have an influence [on your brain chemistry] by changing your behaviors?"

CC: "Yes, for instance, if you can have some things in your day that you tend to feel better after you do them- just small things. Like, you used to work out, and you used to work out two hours at a time. But lately you've been so depressed you don't do that. So, we might suggest, 'How about walking around the block?' Or if you haven't been interacting with anybody, how about a five minute conversation? Not huge pieces of behavior, but some small pieces can be helpful."

NS: "It was interesting, when I got into the elevator to come up here, there was a sign that said, 'Freedom your fear!' It was advertising an application for your cell phone, that if you pressed this button on your phone it would instantly call the security or the cops..."

CC: "Oh yeah, right, it calls the security... I think we've somehow gotten into this funny state where we feel that 'I shouldn't feel fear or anxious'. I think-"

NS: "-That people 'get afraid of fear'?"

CC: "Yes. You have to ask yourself, 'What is this fear trying to tell me?' It is part of our natural reptile brain to feel fear. So, there is this part of it that's called 'affective learning'. So, there is a part of our lives where you are going to worry about some things in your life, and that's normal."

NS: "Fear is a normal thing that keeps us from cutting our finger off when we're chopping up celery. So, we don't want to get rid of all fear..."

CC: "Right"

NS: "It's funny, because I recently read of a case study of a woman who had a damaged amygdala, and she lost that capacity and as a consequence she was constantly injuring herself because she was *totally* fear free. So, that's not what we're after."

CC: "No. But we are after a kind of a relationship with it [fear] that we can work with it and understand.

Maybe, for instance, I'll have some people sit down once a day and write down 'What am I concerned about?', 'What is something that I can actually do something about?', 'Let me make a plan, let me do it,' and 'Let me come back and see how it worked." Go after it in a way that you can do something with it."

NS: "Would you say that when fear is not understood and becomes out of control, that's something we need to fix inside of us?"

CC: "Yes."

NS: "Would you say, given the suggestions you gave, that the way to deal with crippling fear is to use, simply put, our pre-frontal cortex?"

CC: "Mmm, yeah."

NS: "What kind of tools do we have to utilize fear in a positive way rather than be crippled by it?"

CC: "Well, we do have frontal lobes. We have an ability to problem solve, and we have an ability to approach, rather than to avoid. Sometimes we can get into a cycle of just running away from things. So we can learn to slowly approach and have a different relationship with our fear and anxiety that's not so overwhelming.

But people can get into a fear response- just with the fear. It can get into a vicious cycle. So I think some of the techniques are to break that down into sizable bits that they can really work with, and also to not avoid it."

NS: "You have a reactive brain and you have your frontal lobes. And you have your emotions which are like a fulcrum in the middle between the two. Would it be fair to say that if you are crippled by negative emotions that you are being controlled too much by your reactive reptile brain and you are not utilizing enough of your frontal lobes skills of logic, and higher intuition, and communication, and social interaction?"

CC: "That's an interesting way to think about it. Yes, I don't think that's an unfair way to see it.

But I also think that sometimes you just have to sit with all of those emotions and figure out what that experience was about, and validate that experience also before you can just say, 'Oh, my frontal lobes weren't doing what they needed to be doing.' So sometimes you just get clogged with emotions that never got processed, because sometimes it's hard to just sit with those really painful emotions. You've got to sit with what is going on for a while."

NS: "You've got to ask yourself, 'What's going on here' first..."

CC: "Right."

NS: "We do have these frontal lobes that we can use to analyze and get a global view of things. We can see where we've been- we can time travel and we can go into a possible future and remember what we see and bring it back to the present.

And so I thought, the frontal lobes is like being able to have a bird's-eye view above a maze. You can remove yourself from being stuck right in one place in the maze, and you can fly into the air and say, 'Oh! That's where I am! I'm trying to get from A to B- I can see my relationship to the maze, and other mice in the maze...'

Do you think that's an accurate metaphor about the function of our frontal lobes?"

CC: "Sure. But I wouldn't be judgmental either of our reaction. That gut reptile reaction is there for a reason, and I think we also need to respect that.

I think that sometimes we need to just stop and hold tight. Sometimes we have a reaction that we have to do something real fast. So just remember, 'Yeah, it might be an old reaction, it might be stronger than we like', but its okay to just sit with that too. Just try to get some information, and try not to push the process too fast."

NS: "And just try to understand what's going on. I like that. It reminds me of the Chinese Handcuffs- the harder you pull..."

CC: "The worse it gets. So just hang tight, you'll get information, the energy behind it will settle down, and then you can work those two together- the older reptile brain and the frontal lobes."

NS: "You'll get the equilibrium, because you need both."

CC: "You need both..."

NS: "Otherwise we would have discarded our reptile brain sixty-five million years ago."

CC: "It's a real important component of the brain."

NS: "But not to be ruled by that part of the brain-"

CC: "No, you need to integrate them together, the frontal lobes and the reptile brain- and figure that out."

Karl Teariki
Brain Education Program Designer and Facilitator

Karl is a native of New Zealand who designed and ran an 18 month program across his country teaching young people (often troubled teens) basic brain physiology and function, as well as helping them to learn and invent their own methods for controlling their amygdala.

NS: "You started your own program in New Zealand showing kids how to control their own brain beyond what they thought was possible. Can you describe what that was?"

KT: "I worked for a trust that provides social services to young teenagers who have fallen through the cracks of the educational system. I applied for funding to run a course on brain education and behavior modification. We received enough funding to run it for three six-month blocks.

I toured it around the country at a number of educational providers. Some of them at high schools, some at primary schools. The students were many teenagers, all low literacy, who had a hard time with the educational system. We had some older students as well, and some people who were in mental health assistance programs. There were three groups of teenagers, and two groups of what they call mental health consumers here.

We covered basic brain anatomy and function, we learned how animal behavior related to human behavior, we watched video that showed how different parts of the human brain related to different behaviors- mob behavior, aggressive behavior, we looked at what else the human brain can do with the frontal lobes, and then we found where our amygdala was.

The students came up with their own ways to control their amygdala, because they learned what it was and what it did. They learned what emotion was and how it connected to the amygdala, and they started playing with ways that they could turn on their frontal lobes."

NS: "What did they come up with?"

KT: "I gave them a lot of options. They used keys, light switches clicking forwards, or feathers."

NS: "So primarily they visualized images and imagination to do this?"

KT: "Yeah. I gave them a whole lot to choose from."

NS: "What kind of results did you observe, both short and long term?"

KT: "We did half-day workshops for the younger students, and some of them reported visual things- some didn't experience anything, but others got quite a surprise because they weren't expecting anything to happen at all.

Some would see colored lights with their eyes closed when they tickled their amygdala, others would get goose bumps. Some would feel something inside their head, others couldn't describe exactly what it was and might say, 'I don't know what that was, but it was cool!'"

NS: "So, was this a one-shot workshop that you would bring to different facilities?"

KT: "It was actually two different things. One was like a crash course, just a morning or an afternoon workshop. The other thing was a regular group, where I would see them once a week for an hour or two hours, and we would meet about twenty times over the course of six months."

NS: "What did you notice about the groups that you saw over the long term?"

KT: "There were quite a few examples- and this was even before the funding. A guy was in the lobby of the place we were holding a group, and he was yelling and screaming. I went to see what was going on, and it was because someone had said that someone else had said, 'Yada yada yada', and his daughter was very upset about it, and he was getting ready to get in a physical fight. Typical teenage stuff.

So we calmed him down and we defused the situation.

The girls who were taking the brain course realized that they had handled it differently than they would have previous. One girl said, 'If this had happened before, I would have punched that guy!' And we all looked at each other and laughed."

NS: "And what the participants in the group observed was that they had defused the situation, and they associated that with what they had learned about their brain?"

KT: "Yes, and it surprised them too. They didn't expect it.

Through the course we would try different things. We spent the first few sessions learning about the brain, and the triune brain, and learned about the amygdala- then we tried a little bit of art, and a little music and writing, watched some videos, a bit of reading.

Once we recorded a song on the laptop computer, with all the different parts and harmonies. And the kids would complain, 'Oh, I can't sing, I can't sing.' But by the time we finished, they had something they were really proud of. We also had them do some art.

Another example was from a fifteen year old boy who couldn't read well. He didn't have many goals or dreams. When I first saw him, he wouldn't try anything, he was afraid that he would get things wrong.

At the end of the course, this same fifteen year old kid said, "I think I enjoy working in the kitchen, and I'm going to try and get an apprenticeship to learn cooking.' This was really a surprise, because normally this boy had a scowl on his face. So with the help of a tutor, he actually got an apprenticeship. I saw him a few weeks after the course finished and he was quite happy and proud of himself."

NS: "Did you continue to review the brain aspect through the program?"

KT: "Yes, we did a little bit of that every week. For them, it was nice to know what is going on up here [in the brain], and that they weren't at the mercy of temporary things."

NS: "What you're saying is that they understood how their emotions were connected to things, and that they had a tool to understand themselves better."

KT: "Yeah. Everyone who took the course said they found it easy to understand, not too much information at one time.

Another young girl was interested in social services, which became apparent about half way through the course. She went on to get a national certificate in social services. For someone who had trouble reading without even any school qualifications, this was a very big thing for her. She went on to study even further as well."

NS: "Did you give thought to the idea that if you hadn't included brain information and amygdala tickling exercises, that you would have still seen positive results? Or do you see the brain and amygdala education definitely part of the success that you had?"

KT: "I do believe that it made a big difference to have the brain information included, otherwise I would have just had a course that was a bit of art and music, and reading and writing. Obviously it was the brain element that gave them something to understand themselves with. We did it all the time, often starting out watch videos of people and animals and their behavior, and we would have a discussion of what was going on inside these animals brains, and these people's heads."

NS: "So this sounds like a well rounded course, so the students would understand things in some depth.

So what personal experiences have you had that motivated you to create this course?"

KT: "With me, I noticed that tickling my amygdala generated creative thoughts whenever I needed them. I knew it was very useful for me, and I wanted to share this with people who might have needed it more than some others."

NS: "How did you hear about this technique of tickling the amygdala?"

KT: "I had watched a television program that at one point said, 'Your brain can make more powerful drugs than anything humans can make- and I thought, 'What?!' I started thinking about the chemicals that our brain makes, and I thought, 'There must be a switch somewhere in our brain that I flip that will open something up.

Then I did a web search, and I came across your web site. And I read a little bit, then I came back later and read a little bit more, until I started to understand what you were talking about."

NS: "What happened in your own experience that kept you going?"

KT: "I read one of the posts about amygdala tickling, and when I did it I felt a little bubble and buzz inside my head, and from then I realized this was real."

NS: "You knew I wasn't making this stuff up (laughs) You know that I wasn't just some nut- because it actually did something when you tried it."

KT: "Yes. (laughs) That's what I tell someone who tries it and feels anything that goes on in their head, even a little thing. 'You just proved that this is real.' And even the one's that don't feel anything at first, it just takes a bit of practice, like riding a bike."

NS: "What else?"

KT: "Things that might have gotten me really worried, little situations in life that would have stopped me in my stride and have me freak out- if my brain goes into anxiety and I can't think properly- I could just relax and stay calm. I would think, "I can just use my brain to do this, and not just my reactions.' Those negative things really decreased."

NS: "You could identify when you were clicking back into these negative things, and you could exercise some control and minimize that."

KT: "Yes, and that was the feedback that I got from a lot of the students, that they had some control- They might still get upset, but from that point of calmness, it was a different way of looking at it. The things that would have upset them to, maybe a '10' last week, this week it would only upset them to a '5'. And then it was easier for them to deal with it."

NS: "Some people keep a written record of their emotions, and people report a gradual increase towards the positive and a reduction of the really negative valleys."

KT: "Also the perception of things has changed for myself, I'm not so taken in by illusions and the smoke screens, like so many people are. The brainwashing, the politics, and the media, and the corporations."

NS: "So do you continue to visualize and tickle your amygdala?"

KT: "Yeah, regularly. I do it any time, whenever it occurs to me."

NS: "Well, I congratulation you on the superlative job you've done in helping other people learn about what goes on inside their own head. You are really one of the pioneers- and we're just at the beginning. I think we're going to see this just take off in the next few years, and you've got an important story to tell."

KT: "The planet needs to know how human brains work. I sometimes think that our descendents will cause this 'The Stupid Age'." (laughs)

NS: "Well, let's hope this is the beginning of "The Brain Age.""

KT: "Yes! Lovely!"

AMYGDALA TICKLE #41- "PIES Graph "

Eat a slice of your favorite pie.

Now that you're in a good mood, get out a sheet of lined paper and a sheet of graph paper (to plot a graph) and start keeping a record of your PIES- Physical, Intellectual, Emotional, and Spiritual points: -10 is suicide (and if you hit that, don't bother with this any more) and +10 is total Nirvana.

Use a different color for each piece of the PIES on the graph.

See how your PIES points change after you begin tickling your amygdala- after one month. Then try six months.

Then have another piece of pie. With whipped cream.

Mike McCartney
Public School Music and Math Teacher, Composer, Musician

Mike McCartney's was recognized for decades as being one of the most capable and well liked music teachers in his Colorado public school district. He consistently produced great music students and absolutely remarkable sounding ensembles. Mike was my student teaching supervising teacher when I earned my state teaching certificate in college.

He studied with internationally recognized composers Normand Lockwood and Cicil Effinger, and additionally performed with, wrote for, and arranged music for his own commercially successful music group.

MM: "Early in my life I wanted to be a teacher and a performer. I didn't choose education as a backup if I didn't make it as a performer, which is certainly the case many times with music teachers."

NS: "What was your motivation to teach?"

MM: "When I was in high school I had a hard time with math- I was taking the "dummy" math classes. All my buddies were in the upper math classes, and I wanted to join them. My dad inspired me because he only had an eighth grade education, and he was starting a career in engineering, and was taking calculus.

When I took a test to get into the higher classes, I didn't do well and the head of the department told me, 'Oh, you'll always do bad.' I thought, 'What a terrible way to motivate students!'

So I thought right then, 'That's not right, I want to be a teacher and motivate students instead of discouraging them.' As a result, I ended up taking math as a minor at the University of Denver, and began my career teaching both math and music."

NS: "I can remember in high school a few great teachers who made them. Can you remember any teachers like that?"

MM: "Oh, absolutely. I'm very grateful to an English teacher I had in junior high who really taught me how to write, a Latin teacher, various music teachers."

NS: "What was different about these people and why you remember them?"

MM: "It was a combination of their personality and that they had the content and could project it and make it interesting."

NS: "I had a teacher named Ted Tsumura, who was teacher of the year. When you went to his class it was fun. It wasn't just having facts crammed into your ear. He went out of his way to make it a fun adventure. My physics teachers, Mr. Robinson and Mr. Keefe- heck! I still remember their names forty years later- that's how good they were. They understood things from the perspective of the students. You couldn't wait to get to that class because it was like being on a game show. All the other teachers, for them teaching was just a job."

MM: "The first job offer I received was as a vocal choir teacher, which I wasn't trained for. I wasn't a vocalist, and I tried to explain that to the people who wanted to hire me. They wanted a diversion from their regular classes and there weren't any math or instrumental music openings, so I reluctantly took the job.

It started out pretty rough, and I was ready to resign even before the school year was up. But the head of the districts music program encouraged me and said, 'Aw, it's no big deal. An instrumental job will open up eventually.'

But by the second year, I could play piano better, and I started doing some new things the kids hadn't seen before, new kinds of pop music and show tunes. I also started doing talent shows, kind of like American Idol is now. Before long, I was really enjoying it."

NS: "It sounds like you were tuning into what the kids enjoyed, and tried to make it fun."

MM: "Success breeds success. Once I got to the point where things were going pretty well, I could relax and be more myself instead of being a disciplinarian. Then I got into the elementary instrumental music program, and I was a traveling teacher and went to all different areas of town. They mixed it up so you might have a school in an affluent area, and then you would also be in the poorer sections of town where the school would be under the I-70 viaduct. I really enjoyed the variety."

NS: "What were the hard challenges you faced?"

MM: "Time. You only had two twenty-five minute periods of time each week with each class. Try to get them ready to put on a Christmas concert, and then another in the spring."

NS: "A lot of these kids come in and they start out and they can't play a note…"

MM: "That's right."

NS: "I remember my first semester teaching in the public schools, and it was tremendously difficult, as you come in more as a babysitter than a music director. But you spoiled me, because even your elementary school bands were incredible!"

MM: "I didn't take planning periods. I used every minute to spend with the kids. I split up the classes so I could concentrate more on specific instruments and have each instrument in a separate section, and I had great success doing that."

NS: "So you didn't lump everybody together."

MM: "No I didn't. I think I was the only one who did that kind of thing at the time. It was a lot of work, but the results were good. I started a lot of double reed players, oboe and bassoon, which were just sitting on the shelves in the junior high schools. But I had to teach those kids separately, because you can't teach the clarinets and the double reed instruments at the same time, it just wouldn't work."

NS: "You know I did substitute teaching for a year, and that was really tough. I did everything I could to keep order in the classroom. I did everything I could to keep the kids from exploding, and eventually even brought Mad Magazines into the class for the kids to read one day. Of course when the principal found out she told me, 'Don't ever do that again!' And I didn't. I quit. I figured if it was down to this, public school teaching was not the job for me."

MM: "Substitute teaching has to be one of the wickedest jobs of all time."

NS: "And substitute *band* teaching, it's not just disorder, it's *loud* and disorderly, kids with instruments as weapons. It was a war zone. I thought about buying an old army helmet and taking it with me to work.

So, how long did you teach altogether?'

NN: "Thirty-five years."

NS: "What is your take on the problems of public school education?"

MM: "I saw a lot of change from the first years and the last. Parent involvement diminished, and parents have abdicated their responsibilities to the teachers much more than before, because of various things, family situations, jobs.

Also, you have all of this talk of education reform, and all of the people instigating that are politicians and administrators, most of whom haven't taught a single day in their life. They really don't understand what is going on. I'll get in trouble saying that, but I don't care.

We've come to the point where we make judgments based on the results of standardized tests. What does that produce? Teachers that teach to the test. We spent so much time just preparing kids for the test, it's not good."

NS: "It's like sending kids to a concentration camp."

MM: "Oh it is. So as a result we rate schools according to how the kids do on the tests. But standardized tests are never the complete picture of what a student or a teacher can do."

NS: "Look at who we value in society, the movers and shakers. How well did these people do on standardized tests? (laughs) How did Mozart and Beethoven do? How did Einstein do? What about our current media hero, Steve Jobs?"

MM: "Steve Jobs dropped out of college."

NS: "I guess if you're turning out worker ants or obedient worker bees... Aren't we supposed to value individuality and innovation here in America? How does standardized testing have anything to do with that?"

MM: "You're right. You'll have to ask the legislators about that, and they don't even look at half of the picture. They say, 'Just give us some hard core evidence' and 'Education has to operate like a *business*'."

NS: "You've also been a musical composer most of your life, tell me about that."

MM: "I have a master's degree in trumpet performance, but probably spent more time studying composition. I studied and did a lot of work with composer Normand Lockwood [Prix de Rome 1929, Guggenheim Fellowship 1943, 1944], who was not a regimented guy, just the opposite of we've been describing. I also studied with Cecil Effinger [Prolific American composer, who also invented the musical typewriter].

Composers tend to be very creative people, so they don't regiment you. I'd go to his house. Both of these people were modernists, and didn't like to be pigeonholed into any one thing."

NS: "I discovered the hard way how impossible it is to get your large ensemble works played by any orchestra, so I just put together my own group and recorded it."

MM: "That's what Phillip Glass did, he just created his own synthesizer ensemble. As he was coming up in the composing world, he saw what he was up against and just said 'the heck with this' and put his own band together, recorded, toured. He's made the big time just by building on his own ensembles."

NS: "If you were to give someone some advice in their career, what would you tell somebody?"

MM: "You have to set your sights and work. Don't let failures get in the way. Embrace the failure and learn from it. You just have to work at it.

Sometimes, you just plow through something at first so you get a feel for it. That's important to consider, especially when working with young people- they won't tolerate stopping every minute to get something perfect.

But generally the way I taught and the way I worked was to break things down into its smallest component. You master a passage, feel good about it to motivate yourself, then go from there. You perfect that small piece and then bring it all together."

Paul Kashman
Independent Newspaper Owner and Chief Editor

NS: "When did you begin your newspaper?"

PK: "Thirty-three years ago, in October 1978. It covers central, south-central, and near south-east Denver.

The newspaper was started before there were any community papers for the area, and I was hired as a thirty-one year old salesman to sell ads for the paper. I remember my first month, I made seventeen dollars commission, and I wanted to quit.

But I hung out, and at one point when someone didn't hand in an article, I was asked to fill in. Another time, someone was needed to take a picture, and I had a camera, and so on. About three years later, the woman who began the paper wanted to start a family, so I bought the paper from her for some ridiculously paltry amount of money. And the rest, as they say, is history.

I just describe it as everybody's home town newspaper, except that my standard line is, and I believe it, that it's the best you've ever read. That's because I have some great people who are involved with the paper, and I always have had that."

NS: "So as a salesman and as a businessman trying to get people to respond to you, for example, trying to get people to place ads or perhaps trying to get an honest opinion from a politician that you are talking to- how do you get past that defense system that people automatically put up with their reptile brain? What's your strategy?"

PK: "In both cases- ad sales and communicating with a politician- about the only technique or strategy is to be persistent and to keep trying to open up a dialog. Sometimes that works, and sometimes it doesn't.

With the ad people, with some people they've just had bad experiences with others and they don't trust you. You think of the car salesman image- and all you can do is try to put their mind at ease and let them know that you're trying to help them. Sometimes you can establish a dialog- but then there are those who are simply impenetrable.

And I find the same thing with politicians. You can talk to some and you get some exchange of ideas- but then there are some whom no matter what you do you cannot get them off their message. The Reptilian Brain, that's all there is, that's all you can get them to do."

NS: "We've all run across sales people who come at you like, "Hey! Hey! Let me tell you about this! And you gotta' do this! Hey! Hey! Hey!"

PK: "I've never been a fighter, I'm not a tough guy. I've never gotten what I needed by beating people up. I got what I needed by negotiating my way into positive outcomes."

NS: "You're a diplomat as opposed to a boxer, and that's worked for you. I get a sense reading your columns that you look at issues in a very logical manner. By contrast, I read some columns in our city's big corporate paper, and they come across as very one-sided with about as much thought as a fist fight."

PK: "People come in to me, and they can be very passionate about an issue or a person and I'll listen and want to get to the bottom of it. For example, they'll slam their fists down on my desks and complain about this son-of-a-b named Neil Slade. So, I'll call you up and say, 'Mr. Slade, why did you do this?' And invariably, The Mr. Slade or whatever the situation or whoever it turns out to be gets a big smile and says, 'Yeah, well it's not like that'. And then you hear the other side of the story with equal passion.

Usually, there's no evil going on. It's just different perceptions, and people threatened in different ways, and so on. There are generally two sides to all stories."

AMYGDALA TICKLE #43- "Opposites Attract Brains"
When you find yourself coming to a "Definitely *Definite*" conclusion- Hold on just a wee second before you jump to your 'gums'. Look at the opposite possibility.

The realization of the Yin-Yangness of the universe will never fail to tickle your consciousness, as well as make you a supremely perceptive person. If one thing exists, somewhere, so does its twin.

Robert Reginelli Sr.
Stockbroker

I quickly became close friends with Robert Reginelli and his family when I began teaching his son Bobby weekly guitar lessons over a dozen years ago. The family lives in a historic Denver home that had it's beginnings during the gold rush days in early Colorado history. The story of our unusual adventures together exploring music and the brain are covered extensively in a previous book of mine, *The Book of Wands* (2010).

NS: "Tell me about your parents. Were they wealthy?"

RR: "No. They were probably below middle income. Dad never made any big money, we lived week to week. For Christmas, mom went and worked at the Rexall Drug Store to make extra for presents. That was in the late 50's. They made ends meet. We were never poor, but things were limited."

NS: "What kind of things do you remember best growing up?"

RR: "We'd go fishing and camping. I had a lot going on. I was active in high school. I was kind of a big man on campus and an athlete: Track, wrestling, and cross country. I felt like I had to do it to get the girls. Then after high school I went to Germany as an exchange student.

I came back home and went to the business school at the University of Denver and pursued the money that we never had as a kid. I never really hit it big, but I've made a good living."

NS: "Your motivation?"

RR: "I wanted the 'finer' things in life. When you don't have money growing up, you think that money is what you need."

NS: "The grass is always greener…"

RR: "I wanted to have a nice car, and a nice house. I didn't know how to do it, but Dad helped me through college and gave me four-hundred dollars each school quarter. I earned the rest working at Safeway."

NS: "So let's skip ahead. So, you got the nice car, you got the nice house. What's your perspective from here?"

RR: "Now I want to enjoy the nice car and the nice house, and not have to keep working so hard! (laughs) I want to enjoy what I've worked for, I want to read those books I haven't read. I want to put my feet up and kick back a little bit."

NS: "Did getting those things come with a price you didn't anticipate?"

RR: "Oh- you know, it is what it is. Raising kids was tough at times, but there are good memories. I've had much better than many people. But what I've learned is to respect everybody and not to look down on anyone.

When I was trying to come up the ladder, I was pushing people aside and that I don't do any more."

NS: "How about that struggle in climbing the ladder?"

RR: "Well, you're trying to make the money and pay the bills, and it depends on the kind of person you are. I'm a game player, a Type-A. You play the game and that's a part of it. It's the conquest. Once you've achieved it, you lose interest in it and then you go on to the next thing. Once you get it, you realize, 'I've got it now, so what? Now what do I do?'

There are those kinds of people with seven boats, and four houses, and all that kind of stuff. Other people realize that one house is enough."

NS: "You're in a business where you must know some people who are just loaded. Can you make some general observations? I'm looking at this myth of 'More', and it doesn't matter what it is; money, or cars, or houses, status, women, or artistic recognition- it's anything. There is this persistent illusion that 'If I have *more* of anything, then I'll have more happiness and more fulfillment."

RR: "But you realize that after a while that it isn't what it's cracked up to be sometimes.

But it depends on your personality. Some people would rather just let it slide and not work to get whatever it is they dream of. Other people have times in their life where they are driven."

NS: "That's funny, because I know both those kinds of people. Some people I know are content to just sit and do nothing but talk for hours and days on end without doing much of anything else."

RR: "I know. I know enough of those people. But, they're their own person. You respect them and let them go their own way. They'll get to where ever they want to be- or maybe not. You can't live their lives for them."

NS: "What you're saying is that every person is going to determine for themselves what it is that makes their own life worth living."

RR: "Yeah. You have to respect other people, and maybe encourage them. I always try to encourage people, everybody. I'll say, 'You can do it! Maybe you can do it a little better if you want to. Go for it!'

I'm rooting for everybody all of the time. If somebody is making an effort, if you're trying, I'm for you."

NS: "You've seen your own kids, your four boys run into problems of one sort or another…"

RR: "…And we've helped them all. But you have to remember, if you help somebody too much, they won't help themselves. I've tried to keep my kids humble, to strive for things. If you try to do everything for somebody, that's not good. People have got to do it on their own."

NS: "The story of teaching someone how to fish instead of just giving someone all the fish-"

RR: "Yeah. I think my kids are pretty frugal, they never got the silver spoon."

NS: "I know, I've seen them work their butt off."

RR: "My son Bobby, that's me all over again. He's got an affinity for writing, and he could really succeed if he would put it to paper and put the time into it."

NS: "Ah, but the life of an artist or a writer- that's a big gamble."

RR: "But there's always an end to the road, so you might as well enjoy what you are doing."

NS: "Here's a question- You're standing on the launching pad of a rocket to Mars. It's your last day on Earth. What are your parting words to humanity?"

RR: "Are they going to listen to me?"

NS: (laughs) "For argument's sake, let's say yes…"

RR: "Ask yourself, 'What can I do to help?' And I would say, 'Be the best person you can be, do right by everybody else.

A lot of people just do the job. They just want to get the job finished so they can go on to the next thing. For a lot of people the job is just a job, it doesn't mean anything to them, it's just a means to get the money to do the things that you're really interested in.

I say, do the *best* job you can do, what ever that is."

Sean Danato
B.A., History and Psychology

NS: "I wanted to talk with someone who was relatively new to the idea of amygdala tickling, and you had sent me a couple of emails. You seemed like a good person to discuss this with. Can you tell me a bit about yourself?"

SD: "I graduated from Dominican College, and I studied psychology and history."

NS: "How did you hear about brain self-control?"

SD: "I saw a YouTube video that caught my interest. Then I read about it, and that was it. What I read I thought it was very cool. I can do it with visualizing the feather, but I also like using smells, incense, that sort of thing."

NS: "Oh, okay. That works because your olfactory nerves plug right into your amygdala."

SD: "I'm very sensitive to smells and that method. I like incense- vanilla has positive effects on me."

NS: "Can you describe the effect on your brain?"

SD: "To me it feels like a switch. I switch from a state of monotony to something better. It helps me to get away from anger and frustration, and to calm down. That's also what I do when I visualize tickling my amygdala, when I sit and get out of a negative feedback loop, of blame or something like that."

NS: "Do you try it outside of your home when you might be in a stressful situation?"

SD: "I have. Sometimes it might be hard for me to visualize it vividly."

NS: "Have you tried using a physical prompt, like putting a feather in your pocket or on your car dashboard- or, one other thing you can do is get some almonds and leave these almonds around on your desk or whatever..."

SD: "I haven't done that, but that sounds like a good idea..."

NS: "Just seeing an almond around will remind you that you can be clicked backward or you can tickle forward- for example if you're in a traffic jam and you're annoyed because someone cut you off. You'll see that almond and you can say, 'Okay, I can click backward into road-rage, or I can tickle forward and listen to some music or something positive, or just click forward and blow it off. They're also handy if you get stuck in traffic and you haven't had lunch."

SD: "Also another thing that I've found helpful is just making myself laugh. I don't know if you've done that..."

NS: "Oh, of course... Humor occurs in your frontal lobes. That's an actual fact. People with a lobotomy don't chuckle much. Laughter will tickle your amygdala. If you watch a funny movie or read a funny book, that'll do the trick."

SD: "The funny thing I noticed was that if I just force myself to laugh at the situation it will change, even if I don't find it funny initially."

NS: "In Japan, they have laugh therapy classes. I saw this in a Michael Palin travel DVD. He was going through Japan, and there was a group of people who got together every week, and all they did was laugh and go 'Hahahahahahah~!'"

SD: (laughs) "It's such a social bonding thing that people will just laugh if someone else is laughing."

NS: "And they laughed at nothing! They didn't tell jokes, they just made themselves laugh! And it had the same physiological effect as if they had been laughing from an external source. But the effect was the same, all these positive effects."

SD: "People have a lot more control than they think. But no one has told them that they have control. It's not on TV. It's definitely not in the educational system. You must do what you are told- it's not about your passion."

NS: "Any other thoughts about how you've used tickling your amygdala?"

SD: "I'm a musician and play the guitar. I've written some good things clicked forward into my frontal lobes, so to speak. I've deliberately used my guitar as a tool to do that, as a ritual to charge my frontal lobes. I play my guitar and that tickles my amygdala and allows me to write music."

AMYGDALA TICKLE #44- "Musical Mindfulness"

If you have a musical instrument around the house- don't just let it collect dust. Strum it, pluck it, bang on it, blow on it.

Kids who play an instrument do better in all their subjects. Adults who play an instrument activate both halves of their frontal lobes.

If you sound bad- plug in the headphones, turn down the volume, or go lock yourself in a closet. You can safely tickle your amygdala and still keep the neighbors from killing you.

Shirley Kenneally
Recording Studio Owner

NS: "I met you at your recording studio over thirty years ago, when I began recording my music albums. For the sake of our interview here, explain how you got into that business?"

SK: "Ken, my husband was very interested in music, and played the guitar..."

NS: "He was a captain for United Airlines, a pilot..."

SK: "Yes. We had always had parties around music, with amateur musicians, although they were always very good amateur musicians. That was our social activity. We would record those parties on a little cassette player, just for fun.

It branched out to be more than that when the four-track reel-to-reel tape recorder came out for the general public. Back then in 1976, there weren't really any studios that were affordable for musicians to record in. This was long before the days of home studios like we have now.

When we decided to put our own studio together, it just grew out of our hobby. At first we put the control room in our bedroom, and cut a hole in the living room wall and put in a glass window..."

NS: "That's a pretty serious hobby!" (laughs)

SK: "We would go out and hear music and then invite musicians we liked to come back to our house, because we were so excited about our new equipment. We got a mixing board, we got microphones, we got cables and it just continued to grow with more and more equipment that you need for a recording studio. We started out recording musicians for free."

NS: "For a lot of us it was like school, because you let us learn how to do it and operate the equipment. It was so reasonably priced, we could spend time learning how to do things right."

SK: "In our home it was never more than $5 an hour. Then when we moved it to the other building, it was first $10 and hour, and eventually it got to be $15 an hour."

NS: "Even so, that's still extraordinary. It was essentially non-profit... you spent more than you ever made."

SK: "We just invested it in the equipment. There was one year where we did break even. But the studio was really busy every day of the week, all hours of the day. But Ken and I never paid ourselves for what we did there."

NS: "It is an interesting business model, because you weren't concerned with the monetary profit, but rather the effect on the community, and the relationships that you built, and what you contributed to the arts and the music in Denver. What an incredible gift to people- like me- who would have had that opportunity in any other way, we just couldn't have afforded it."

SK: "And it was a gift to us all. We had so much fun- and there was a real need in the community at that time, because there wasn't anything available at all for the average musician. It provided a service that was needed by artistic people and musicians."

NS: "You have a key interest in art as well that I know about, and you are particularly keen on simple expression, in the way that some art says a lot with some very simple design."

SK: "If an artist is really good, they bring the unseen into the visual."

NS: "What do you mean by 'the unseen'?"

SK: "The invisible- the spirit. So when I see what the artist has brought up, I understand or see something that I wouldn't have understood before."

NS: "That makes me think about how when we look into the world, we are usually just seeing the skin of the apple. But underneath the skin there is all of this other stuff, the supportive material, and the fruit."

SK: "And the essence.

When Fred [Poindexter] paints, he doesn't worry about all the microscopic detail stuff, but he gets the essence or the soul part of something, and I marvel at that. Boy, when he's right on, there's a response that you understand- it's a truth. There's a truth there that is revealed in the art."

NS: "What do you think is the most important thing we can know in this life?"

SK: "Shakespeare said it: 'To thine own self be true.' Know yourself, and not only know it, but be true to it. Because I think a lot of people lead fake lives. You've got to go by your own inner instincts, and trust it. Trust in your intuition."

NS: "But, how do you think you know yourself?"

SK: "You have to be still. You have to have quiet time. Go for a walk in the woods- something that is an interior thing…"

NS: "…That allows you to reflect without any distraction?"

SK: "So you're not influenced by anything, by any outside events. Then you bring that in the world. You don't stay there, but you have to have that time to tap into it. I've learned that if something doesn't feel right, to pay attention to it– because my intuition is probably correct. But you know, you never figure out totally who you are, because you're always changing.

And I also feel like you have to be careful what images you feed yourself, because it all feeds into your psyche."

NS: "Images are food for your brain?"

SK: "Yes. You're always being affected by what you see and hear, and its all cooking and creating who you are. Even the little things, all the time. You're being made new every day by what comes into you."

AMYGDALA TICKLE #45- "Amygdaloid Art Appreciation"
Look at some art. Let it quietly wash over and open up your neural pores. Check out library books, go to the art museum, or get an arty DVD. See how-what-where-why great art tickles viewers by the millions over millennium.

Thomas Taylor
Coffee Shop Barista

Thomas works at the counter at my local coffee shop, and I couldn't help but notice that he enjoys his work tremendously. I knew that he would have something essential to add to this story of brain self-control…

NS: "Tell me what you enjoy most about your work here…"

TT: "Initially my favorite thing here was my co-workers- easy to work with. That's initially what drew me in at this place, everyone is really great here. Secondly, it was the craft itself- attempting to pour nice drinks, and somewhat of making an art form out of it."

NS: "The creativity involved?"

TT: "Yes. People who order drip coffee, that's one thing…"

NS: "So you hate serving *me* then…"

TT: (Evil smile) "I drink drip coffee myself, so… But I do enjoy making the craft drinks, once I got good at it. It became one of my favorite things."

NS: "So there is an art to that…"

TT: "Indeed, indeed."

NS: "The [espresso] machine that you use is quite an elaborate and expensive device."

TT: "It is, it is. But it's not going to make designs with the milk- that's all the person who's doing it."

NS: "I remember Lee who used to work here, who's now riding his bicycle on a year long trip from the Artic Circle to Tierra del Fuego [true] - he had a photograph hanging on the wall here, a black and white print of what he used to do with a latte. I thought it was Photoshopped- but it wasn't. It was something he actually did in a coffee cup. It was amazing."

TT: "We try to do things like that as often as we can. He was obviously incredible. But after I became comfortable with the co-workers and the craft, that really allowed me to start getting to know my customers, and then it was completely different. Then it's not just a job any more- I'm here at *the neighborhood hot spot*."

NS: "So tell me, you're a young guy and still figuring out what you want to do beyond your job here- what do you use to help guide you to make decisions as you go along?"

TT: "To be completely honest, my parents.and my family figure in that quite a bit. I'm really good friends with my parents and with my siblings. I talk with them on a daily and weekly basis. A lot of my important decisions I make consulting with them and with my older brother. And also my friends."

NS: "So incorporating the experience of others keeps you from making mistakes that they may have already learned from?"

TT: "Yes. And I feel like that I'm coming around to that point where I can level with myself about a wrong decision that I may have made, and how to deal with that."

NS: "How do you know when you've made a good versus a bad decision?"

TT: "Typically I just know that I should have made a different decision, I just understand it almost immediately."

NS: "Do you think that there's a way to improve your brain's radar so that you'll have a better way to anticipate the right move in the future?"

TT: "It seems like the problems happen when things get too complex, and then I realize I don't need all of that.

So, a goal in life for me is to sift out what I don't need, and make things simpler. To get rid of all of the baloney that truly doesn't matter, and focus on what is important.

One of the ways we deal with a bad or cranky customer here is to use humor. We just try and laugh it off after we're done getting them want they want rather than let it ruin our day. In fact, its something that helps connect us together here as co-workers. We just have this unspoken communication where we realize, 'Oh yeah, that was ridiculous. Hahhah!'"

NS: "That reminds of watching John Cleese, when he's playing the part of a person who absolutely explodes over something that's completely ridiculous. Yeah, it really is funny- the humor in the way some people react to things completely out of proportion. You have to laugh."

AMYGDALA TICKLE #46- "Tickle Tea Time"

Maybe the Mad Hatter was crazy because he just didn't understand how to make a decent cup of tea.

To the Japanese, tea preparation and serving is elevated to a ceremony and high art of deep contemplation of the meaning of life. Create your own ceremony- with the tea or coffee of your preference.

Glenda Heath
Massage Therapist, Yoga and Aerobics Instructor

Glenda has been an aerobics teacher, yoga instructor, and massage therapist for over thirty-five years.

NS: "You began at the brain lab when I was there, it must have been around 1983."

GH: "Yeah- that's just what I was thinking."

NS: "So, you learned how to tickle your amygdala way, way, way back then. Do you remember how you did it when you started out?"

GH: "I just remember the lessons and the homework that I took home. I was keeping a journal and a graph chart that kept a record of my emotions. The bottom of the scale was, like, death or suicide and the top was the biggest happiness you could experience."

NS: "Minus ten to plus ten... A daily record of your emotions. What kind of things did you do to move your graph results from the negative numbers up to the positive numbers?"

GH: "I remember doing the 'Trauma Dramas', the visualizations to get rid of negative things. [see *FL Supercharge*, Chap. 8, Brain Games]."

NS: "If somebody says to you, 'Tickle Your Amygdala', what do you think of?"

GH: "It's funny, just when you said that I felt an inner sensation of flipping something forward. (laughs) I just think of being 'in joy'. That's what brings me there. Being joyful in whatever I'm doing."

NS: "What kinds of things do you do when you're working on someone as a massage therapist to keep yourself feeling that way?"

GH: "When I step into a room to give a massage, when I put my hands on somebody, I step back and ask that I become a conduit of healing, a messenger of healing for the body of that person. Most of my massage happens with my eyes closed. I just let the energy flow through me. I'm a messenger."

NS: "Tell me a little bit about that energy that you're connecting with."

GH: "I call it source energy. It's peaceful and calming and healing energy. When my clients get off of the massage table, you can see an immediate shift (laughs), people respond to it."

NS: It's not just physical manipulation..."

GH: "Oh no. People get it. It's source energy that comes from the planet."

NS: "Last time we talked, you talked about the standing yoga postures, and how you felt that those brought that earth energy into your body. So, this source energy isn't something that's 'up in the clouds'..."

GH: "Oh no, it comes from the center of the planet. And we are part of the living flesh of the planet. If everyone would connect to that source energy, we would all be healed, the planet would be healed."

AMYGDALA TICKLE #47- "Healing Hands"

Imagine your hands are combs, your fingers slightly apart.

Have a friend or partner relax, while you "comb out" negative entropy from their body. Breathe deeply, and move your hands– perhaps an inch away- over their head, neck, torso, arms and legs, in a soothing motion slightly above their body, down and out through their feet. Go slow.

Shake off the stale energy from your hands. Then finish, palms facing downward, sending comforting source energy waves of healing, through you, into them.

Wil Rickards
Outdoor Education Teacher, Mountain Climber

Will Rickards also has an extremely strong connection to the Earth, but in his own way as mountain climber, outdoor educator, and with his work in sustainable land use design.

NS: "Where are you from originally?"

WR: "I was born in a little university town called Bangor, in Wales. I studied Environmental Studies, traveled a bit, got a teaching degree and a got an M.A., worked in outdoor education for twenty-five years, and did some work in sustainable agriculture. I spent some time with Bill Mollison in

Australia working with permaculture and sustainable design. Bill always used to say, 'By myself, I can't do a thing but with one friend I can change the world.'

I started in the UK as a woodwork, metal work, and technology teacher for a little bit. But I was also a climber, a kayaker, and a skier, and eventually realized it was a great arena in which to teach, and for me more important than any other subject in school."

NS: "You were a fairly seriously mountain climber..."

WR: "I started climbing at the age of 15 and was fortunate enough to go to the Alps when I was 16, and ended up climbing all over the world."

NS: "Tell me about that."

WR: "I hung out with people who didn't necessarily get a lot of media attention- they were just good, good climbers. A number of them were famous in other walks of life, perhaps as an artist or a writer, folks like John Redhead, and a few were famous as climbers as well."

NS: "When I think of Wales, I don't think of big mountains..."

WR: "No, but if you're into rock climbing, it is a mecca, and one of the places where it really started. The diversity is incredible; you can climb thousand foot mountain crags, four-hundred foot sea cliffs, little outcroppings..."

NS: "What's been your internal guide that you've used to find your various paths in life?"

WR: "Historically, I've always thought about what makes me happiest. And it's always come down to what I'm contributing and giving, and the feeling that it is something valuable and worthwhile.

At the moment, it's parenting. Before that it was definitely teaching. When I was younger and a little more 'selfish', climbing was great."

NS: "But what I see in you are two polar opposite activities- Mountaineering is a very solitary activity, with few social benefits..."

WR: "Actually, the social benefits of climbing are quite huge. You bond with people in a way that you normally don't do. If you are sharing something that is potentially harmful with someone, then you get to know them and trust them way more than normally. It brings a sense of self-awareness, knowing what you are capable of and also what it means to truly trust someone.

Those are themes that run through what I teach, and what my climbing experience has led me to understand. If I hadn't spent time in dangerous places doing exciting things with people, I wouldn't have had the same understanding as I do now."

NS: "When I look at mountaineering, all I see is this adrenalin rush, and egotistical physical challenge. But what you're saying is that there are things that coming into play beyond that sheer thrill."

WR: "I'm not going to say that all climbers are this way. But the climbers I was attracted to be with were more in the eastern tradition of the old warrior, or a martial arts monk. It is a far more soulful activity than what people from the outside might recognize.

We tend to pamper ourselves in our society and enclose ourselves in a way that is comfortable, and we forget what life might really be like. Any of these exploratory sports where, yes, you're sleeping on the ground, and you don't have the modern amenities of life- sure yes there's discomforts, but some of us crave it because when we don't get this we're separated from our true selves.

This is in a way a real life. If you were living in the same place you do now, three-hundred years ago, you'd be a native riding around on the plains and visiting the mountains. You would know what it was when a storm came through, and when it rained and when the wind blew. You'd know what it was to be hungry to see extreme beauty. And we cut ourselves off from that."

NS: "So you're saying those extreme activities amplify what it means to be alive?"

WR: "Very much so. But that feeling of adventure when you're uncertain of an outcome is something that you can get in any situation, whether it's in your bathtub or on a cliff somewhere. It's a state of mind.

I know that I'm living a compelling life when I jump out of bed in the morning and I see reflected in the people around me that they're as excited to be around me as I am to be living. It's an intensity that sometimes hurts and sometimes is hard, so it's not always a good feeling- but it's a heightened feeling. You know when you are there- time stops. You're totally in the moment.

Mihaly Csikszentmihalyi is a psychologist who talked about 'flow' and optimal living, identifying those optimal moments in life. He worked with a lot with artists and climbers in his research."

NS: "What I think of is in terms of a roller coaster of these up moments and these down moments, but you endure the down moments because you realize that there's a higher vantage point down the road. You endure the discomfort and the temporary failure because sense something beyond that's of greater satisfaction."

WR: "Even those down moments can be satisfying, knowing what you're capable of- knowing that what you can endure is satisfying even though it's not 'enjoyable'. When you think about Shackleton and rescuing all of those folks in the South Pole and what those people went through- and yet I don't think any of them wouldn't have not wanted to go through it. They all recognized that it was a valid human experience. Difficult yes, valid certainly."

NS: "So your focus is not on the level of pain or pleasure, but rather on something of greater value above that. So you enjoy the pleasurable moments, and yet endure the difficult and negative spaces…"

WR: "…and value them. It makes you a richer person. If I have stories to tell, if I am an interesting person, that's what makes life worth living. You are the sum of all of your experiences. The more experiences you have, the more you can be.

But, one of the most interesting people I know is a farmer, and he's rarely left a square mile of land. He's occasionally gone to the local town six miles away, but I don't think he's ever been to London. And yet, each moment for him is growth.

They talk about the nuns who do all of the puzzles, because they recognize that if they don't keep on doing something and growing that life is not worth living as you get older.

As humans we need to grow- but that growth doesn't necessarily have to be outwards or from our stomachs (laughs)- it has to be cerebral. It's the new connections of the neurons that makes us excited."

AMYGDALA TICKLE #48- "Terra Tap"
Connect with terra firma. Grab a walking stick and hit the trail.
Plant your feet on something other than an elevator floor or sidewalk. Don't be a wimp and burn a couple of extra calories.
Let the fresh air refresh your mind. Be illuminated by the natural environment. Get some real thinking done.
"Solvitur Ambulando- it is solved by walking."- St. Augustine

---

Walter Gerash
Attorney

Balance is a well known icon as illustrated by the scales of justice, as well as is in matters of morality, and the forces that shape our society, from within and from without.

Walter Gerash has spent his entire life as an outspoken legal warrior for civil liberties and civil rights with a long list of recognition, life time achievements, and awards. He has fought and won many high profile civil liberty cases, and is credited for having won two landmark decisions in the United States Supreme Court.

NS: "Please tell me some background about your life."

WG: "I was born in 1926 in New York City, raised in a tough neighborhood in the Bronx by working class parents. My parents were immigrants from the Ukraine and Romania."

NS: "I know that you've often been motivated by social and moral principals and issues behind the cases, it's not that you just liked to fight."

WG: "I felt that the depression was restrictive and harmful to American people, and I was always wanting to change society. When I became a lawyer I decided to utilize the law to try and help working people and try to change the system."

NS: "You were very, very successful in court during your career. Can you share some principals that you depended on to guide you in way, so that you knew that you were going in the right direction?"

WG: "The key thing is to get all of the facts of the problem. Then relate the facts to the general background of everything so that you are rooted in the basics."

NS: "So, first an analytical and logical process."

WG: "Yes. Everything arises from a unique situation for each problem. So you can't stereotype what your tactics and your strategy is going to be. In a criminal case and in a civil case, the jury decides the destiny of the case, and it's very important to be able to select the right jury- they determine the outcome of everything."

NS: "When you're picking jurors, what do you look for?"

WG: "Basically, if you're representing minorities, you try and get minorities. Generally, minorities are charged more in criminal cases than anyone else. I was instrumental in making sure that minorities get better representation on juries, and that's what the Supreme Court cases were about.

When I formed my legal partnership, I made sure I had diversity in the people I worked with, and these other members of my legal team looked at things in a different way that I could alone. I strengthened my ability, and I credit that pretty strongly to my success."

NS: "As a lawyer, you've seen the worst in what society is capable of. What so you think is the core root that causes these difficult events and confrontations and conflicts between humans to occur?"

WG: "Basically, I feel that society that tries to squeeze as much profit out of working people as possible, and this drives many people to desperate measures to survive."

NS: "Are you talking about greed?"

WG: "It's not so much that. Look, corporations are dedicated and designed to make profit. And so it turns individuals who are part of that system to be driven to do things that they would not do if they had the same opportunities as better well off people.

The majority of incarcerated people are of minority background. It's not because they are inherently more evil than the middle class white population. It's the nature of society that governs who is exploited, and what that results in. It's the system itself that is to blame."

NS: "We have a very high disparity between corporate CEOs and the people who do the work…"

WG: "But its not that the CEOs are bad people- it's *the system*. That's my position, that the system has to change.

One's ideology stems from the position a person occupies in our class system. What working people are now realizing in these depressed economic times is that they're not being protected. And as people get together, talk about, and organize what has to change- that's what will change things."

NS: "What you seem to be talking about is a group consensus that works to the benefit of many- as opposed to a solitary individual trying to make his life better all by himself."

WG: "Of course. Many animals cooperate with each other, and they get things done. People need to do this as well, especially now. The family, even just a regular family is a sort of communal force. And the family survives overcoming unemployment, divorce, or what have you."

NS: "So are you saying that as a society we need to extend the idea of family beyond what is inside our own four walls?"

WG: "Absolutely. Absolutely."

Erfie Da Werfie and Chloe Da Whoaee
Canines

NS: "So, Chloe, some people say that people are smarter than dogs. What's your opinion on that?"

CDW: "Arf, arf, aaaaarrrrrf."

NS: "Erfie, do you concur?"

EDW: "Woof. Da woof."

NS: What evidence do you two have that might substantiate your claims? Obviously, there is a different ratio of, say, frontal lobes capacity to reptile brain- that applies to all, say Chihuahuas and even Great Danes..."

EDW: "Woofy woof woof."

CDW: "Arf arf arf. Arf arf arf. Arf."

NS: "Okay, well I see how that might be possible..."

EDW: "Woof woof woof."

NS: "Yes. Yes, of course. Alright then... would you please put my sock down? Okay, if we include early wolves, and then if we go as far back to the late Eocene age and the Leptocyons and Eucyons, your species certainly has had a lot more practice. I agree.

But what about happiness? There seems to be a certain ease that, at least most domestic dogs in a typical household, regardless of how many squeaky toys you may or may not have... this innate ability to tickle doggie amygdala under most any circumstance. For example, a cardboard paper towel roll or even a dirty stick in the back yard. It doesn't take much, really.

Chloe, you in particular, there's an exceptional amount of licking and kissing as a result. Why do you think that is?"

CDW: (licks lips) "Slurp."

NS: "Oh, yes, that's obvious. But is it learned behavior? Clearly, you don't communicate that through spoken language- at least not the kind that human beings seem to understand."

CDW: (pant pant pant)

EDW: (scratches behind ear with rear foot) "Arrrggh!"

NS: "Is that why we see so much tail wagging throughout history do you think?"

CDW: "Yip."

287

EDW: "Woofy woof." (wags tail momentarily)

NS: "Erfie- please stop chewing my shoe...

Okay. I would tend to agree. It's certainly only a matter of time before scientists find the direct connection between your amygdala and your tail.

One final question- do you prefer dry dog food, pre-moistened- say out of a can- or a variety that includes vegetables and fruits and the occasional yummy tidbit- and I'm not speaking in excess of course- that say, I might be having myself while watching a movie, provided that it's of a nutritious nature and not junk? Is this an example of cooperative frontal lobes behavior?"

EDW: (grabs furry squirrel toy) "Squeak!"

CDW: "Arf! Arf! Arf!" (grabs other end of toy and commences tug-o-war)

EDW: "Werf werf eeeerrrrrgggggg"

CDW: "Errrrgggg!" (big squeak)

NS: "Okay, well, thank very much for your time. You've both given me a lot to think about. Do you want a cookie now?"

EDW and CDW: "Wowwwy woow ahoooooooo!!"

Julia Lu
Fine Artist, Ph.D. History and Philosophy of Science, J.D., Attorney at Law

Julia's interest in the powers of the human brain came about as a result of trying to understand a number of unexpected and spontaneous paranormal and out-of-body experiences she had begun to have while at the University of Missouri, Columbia School of Law. Given her straight and conventional academic background, these startling experiences were something that came as quite an unanticipated surprise to her.

Since she came to the United States, she has earned a master's degree in Philosophy, a Doctorate degree in The History and Philosophy of Science, and a Law degree and then become a licensed attorney.

Eventually I learned that her real love was drawing and painting. Before long, she decided to move five-hundred miles west to join me in Denver where she could become painter full time. She has been painting exclusively since then. As an artist her output is staggering and is an unambiguous testament to her love of creativity and the visual arts.

NS: "Where do you think your creativity comes from?"

JL: "We were all born with creativity and curiosity and all kinds of talent. But this gets repressed by the experiences that you have; growing up, what your parents tell you, what your teachers tell you, peer pressure, conforming to the rules and conventions of society and codes of behavior, even when you're in kindergarten.

If you can free yourself from that, then you can release great creativity and talent. We want to do the best we can- but what does that mean? I think it means to fully release our potential, to honor our talent and make the world a better place.

How can you do that? Only if you follow your passion, only if you're free of the pressure added by anyone else around you or the society."

NS: "So when you're painting, what are you thinking about? Are you consciously dismissing these inhibitions?"

JL: "When I made a decision to become a painter and to leave my law practice I already went a long way to overcome those prejudices and conventional thinking. If you look at society's point of view, a lawyer is respectable and intelligent. But a painter is always on the edge of society.

If you 'make it', and break through and become big like Picasso, maybe people will honor you..."

NS: "If you become an iconic figure..."

JL: "Yes, otherwise you're just another 'artist'. Everybody is an artist, true, everybody could be, everybody is potentially that.

When I was a lawyer, I was happy to provide a service- but that wasn't my passion, it wasn't where my heart was."

NS: "Let's say we've got a sliding scale from 0 to 5, and on the left you have no emotional payoff from your work. Then on the other side, you've got 5 which represents the most gratifying thing you can think of doing. So, where do you rate your experience as a lawyer, and as a painter?"

JL: "As a lawyer it would be about a 3. As a painter, a full 5.

The difference is this: When I was a lawyer I tried to do the right thing, but it was a constant conflict between my own interest and my client's interest. I tried to help people, and it was very satisfactory especially when someone would tell me, 'You saved my life', or 'You made my life better', and you got paid as well to help others.

I don't mean that a lawyer's job is not meaningful, but for me I find a deeper feeling and satisfaction in painting than in anything else that I've done before."

I remember that you had been telling me to forget about being a lawyer and to just be a painter. The day that was a turning point was when I was sitting in court, I was waiting for my case to be called on. What I was doing was drawing everybody in the court. I was drawing the judge, my fellow lawyers, I was drawing my clients..."

NS: "You were doing artwork while you were in the courtroom as a lawyer??"

JL: "I wasn't thinking, I was getting bored waiting..."

NS: (laughs)

JL: "I began sketching everybody, and I just realized that I had started that since the beginning of law school, always sketching people. And I thought, 'I'm in court, I'm supposed to be totally concentrating on my cases, and I'm sketching...'"

NS: "I've done that teaching music lessons, but that's a bit different than being in court when someone's life is on the life and you're drawing pictures..." (laughs)

JL: "But when I was at home painting, I never thought about my cases. You know? So that said something about where my heart was. And I thought maybe my client would be better off with another lawyer..."

NS: "...'Your honor, I don't know about my client, but take a look at these drawings I just did, and if you're not convinced that he's not guilty I'll eat my license to practice law...'"

JL: "No, no, no. I already prepared the case. But what it meant was that was busy thinking about their shape and their gestures... you know?

So, I thought about my friend Cecelia, who is a lawyer, and she's more experienced than I was."

NS: "So, you're thinking that if you leave, your clients will get a better lawyer, and that you'll be doing them a favor if you quit."

JL "Yes. And she's passionate about practicing law, and she cares about her clients deeply. So, that's what I'm thinking, people can always get another lawyer. But if I don't paint, no one else will create the work that I intend to create. Do you understand?

We're all different- we all have a different philosophy, a different life experience, different soul energy- If you're true to yourself, your painting will be different than anybody else, because you are you!

Remember what Mr. Rogers used to say? 'You're special. There's nobody else on this planet exactly like you.'

I didn't go to art school, I didn't have formal training, I didn't go to workshops, I didn't have teachers- and to some degree that's a disadvantage, because I'm having to learn everything on my own. But on the other hand, it's also an advantage because I'm not bound by any kind of conventional rules."

NS: "You are constantly bouncing around the house like a happy ball most of the time. What do you do when you're confronted with something that's difficult or unexpectedly negative, or circumstances that you prefer not to be in? How do you get around that road block?"

JL: "It's how I see the world. We all live in the same world- but actually we all live in our own world, like what that guy said..."

NS: "Oh, you mean Robert Anton Wilson. He said we all live in our own 'reality tunnel'- or the way that I put it, that each person lives in their own You-Niverse. And that all these You-Niverses intersect 'out there', wherever that is."

JL: "Yeah, like how we're all facing the same outside world, we all perceive the world in different ways, and we all react to it in a different way.

To be happy...look, happiness is a state of mind that has little to do with the environment. Maybe *something* to do with it."

NS: "So, how do you manipulate your state of mind?"

JL: "No, you just have a certain way of thinking, so you can resolve a lot of things. I don't get frustrated by things very easily. I don't get angry or mad..."

NS: "I can testify to that..."

JL: "I don't get sad, because why? I wasn't always like this, but now I'm doing what I love to do and I get to share that with others.

So I get to create beauty and share with others. All those people who bought my paintings are very happy, and I know I'm making more progress every day.

I used to throw away canvases that I didn't like, but something happened that changed my mind. This one day my neighbor had a visitor, a Russian woman, and she knocked on my door. When I answered it she said, 'Is that painting yours?'"

NS: "That was a painting you put outside your apartment door that you were going to throw out?"

JL: "Right. And I said, 'I don't want it.' And she said, 'Can I have it?'

And I thought it was a really bad painting. But I said, 'Sure, go on and take it home.' And she asked me to sign it. I really didn't want to, but I did anyway. As it turned out, she worked for an art gallery owner, and she took the painting home and the gallery owner saw the painting and asked to see more of my work. So he came over to my place, and I had all these paintings sitting around on my floor- and he and the Russian lady took fifty paintings out of fifty-four, and did an entire show at their art gallery."

NS: "Holy cow. And you had no idea what you were doing?"

JL: "He had the Fine Art Gallery of Hannibal, Missouri and sold my paintings for two years. So from that time, I never threw paintings away- because I realized that even though I might not like something, someone else might and I should let them enjoy it."

NS: "This is related to the whole idea of what is 'expected'. You may expect something out of what you might do yourself, and at one point in time you might not appreciate it- but somebody else might.

What I've found with your paintings is that there are some that you do that I really like- but they don't sell. And I'm happy about that because I get to keep looking at those paintings."

JL: "Some paintings you don't think are so good- other people are crazy about them."

NS: "So what you're saying is sometimes you're channeling energy of something that is of little use to you, but you find that magically in some way you've done something that is of use to someone else."

JL: "Also, as an artist you change. So things you do in the past just reflect one stage of your career."

NS: "That reminds of a time in high school when I had to take a ceramics class from a pottery teacher who was visiting from Japan. He spoke almost no English at all. So the assignment was to make a little house out of clay. So I built this very rudimentary block house, very simple, four walls, a simple cut-out door and cut-out window. And I just left it in the class shelf and I didn't even care about the assignment, 'Make a house' he told us. So I did, but I hardly thought about it.

A couple of weeks later I saw my house on his desk. And what he had done was to fire my house in the kiln. He had put this multi-colored glaze on the two slabs that made the roof of the house- and it was absolutely beautiful.

All I knew was that I wanted that house back! And this teacher barely got out the words, 'You didn't want it. You forgot about it. It's mine now.' And I pleaded with him, and he was nice enough to let me have it. And my mom still has that clay house in her living room forty years later.

I had rejected my own work, but he showed me that I just hadn't finished it. Or maybe he was saying, 'Two of us together can create something that you didn't quite see all by yourself.'"

JL: "Painting is a bit different, because you paint all by yourself. But you still need input from other people, and I always consider what other people say, although I might have a different opinion."

NS: "What's next for you?"

JL: "I will continue to sell paintings, but I'm not painting to sell. I paint what moves me, what ever inspires me. I don't want to be starving (laughs) but I still want to improve my skill and sharpen my vision."

NS: "Okay, so Julia, one of the things that I'm asking many of the people I'm interviewing for this book is 'How do you tickle *your* amygdala?' Are you familiar with this particular region of the brain?"

JL: "Of course. I've read all of your books."

NS: "I'm not trying to turn this into a commercial-"

JL: "Hey, I translated your book. You need some memory pills." [Julia translated *The Frontal Lobes Supercharge* into Chinese.]

NS: "Oh, yeah, that's right. Haha. Seriously, obviously you know what tickling your amygdala is 'cause you're married to me. So, my question is: What have you noticed? Do you do it? How do you do it? Have you stopped doing it? Why did you stop? What is your relationship to the amygdala, Julia Lu?? Go!"

JL: (laughs) "I can always visualize any part of my body. So I can really see the two little amygdala when I close my eyes. And I can tickle them with an imaginary feather. Very soft, where you can barely touch. It's very pleasant, but not exciting. It's calming- not excitement."

NS: "Well, excitement is something that happens on its own, but tickling is like turning the key on a car so it starts, and then you drive somewhere."

JL: "Yeah. It's more like pleasantly calm and calmly exciting. And if you close your eyes, you can see the fire in your brain. You can see those cells are firing…"

NS: "You see the packets of neurotransmitters going across the synapses…"

JL: "When you think something, you think 'faster, easier', and you wake up all of your creativity somehow, and it makes you happy in the long term.

I like to think pleasant things- I think 'I'm doing this thing I love', I share with others, other people are happy to get my paintings. And also at home, my husband loves me, the dogs love me, my cat loves me…"

NS: "The goldfish love you. Are you saying these are the long term effects of tickling your amygdala?"

JL: "Yeah."

NS: "Like putting a lot of little pennies in the piggy bank, one day there's a lot of money in there."

JL: "Yes. You get used to it and after a while you don't have to close your eyes and it's almost instantly…"

NS: "That's what I explained on the radio from the beginning, you just imaging a little feather and you flick it, and that's it."

JL: "Exactly. And I think that if you keep stimulating it, you set your brain to a default, to a default mode forward. Then it becomes second nature.

And you will feel that pleasantly happy, calmly excited- it becomes your normal state of mind, your default state."

NS: "One comment that I hear from people is that, "It was good for a while when I was first doing it, but now when I do it I don't feel anything.' Did you experience anything like that?"

JL: "No I didn't feel it go away. I feel like it stayed with me and gradually got even better."

NS: "But you also changed what you did. As you moved along your path, you didn't remain stagnant."

JL: "That's because we all have to grow. I accept the fact that I'm not perfect. So I'm not frustrated by my imperfection. Like if I make a bad painting, I always think, 'Tomorrow I'll do better.'"

NS: "How's this related to tickling your amygdala?"

JL: "Because you're always in that mood, in the default mode- in the pleasantly calm, and calmly exciting mood. If I get up, I'm just happy. And if something goes wrong I just say, 'I'll deal with that.'"

NS: "So when something goes wrong, you say, 'I need to tickle this little button in my brain and stay calm and have the energy move forward into my frontal lobes...'"

JL: "Sometimes something happens suddenly fast, and you have to make a decision in a second- like that day Erfie and Chloe ran out the front door, I almost panicked. But I said, 'No. I need to calm down and let my frontal lobe work.' It wasn't like I needed to think a long time, it was just instantly. So I remember that you taught them when you say 'Stay!' they stay.

So when Erfie started to run in the street if I panicked, I would have forgotten what to do, so I calmed down, I took a deep breath and said, 'Stay!' And they both stayed."

NS: "You saved their lives."

JL: "I saved my life too, because if something had happened to them, you would have killed me."

NS: "Well, that should convince anyone. Julia Lu is alive today because she remembered to tickle her amygdala when the dogs ran out the door. And here they are, right here..." [Erfie and Chloe hop on the couch]

JL: "Sometimes when you get frustrated you're in the wrong mode. You know that too."

NS: "Yes. Your emotions tell you that you're not using all of your brain. So at that moment you say to yourself, 'Energy-*forward*!'"

JL: "Paahhhnnnnnnnng!"

AMYGDALA TICKLE #51- "Tickle Reminders"

Make a few little signs, and place it in various places where you live and work to remind yourself to "Don't Panic- Tickle Instead!"

Put notes in the butter box, on your bathroom seat, on the washing machine, on your car's gas cap, on your baseball cap, on your remote control. Any time is Amygdala Tickling Time.

REFERENCES

AMYGDALA TICKLE #52- "Explore"
Explore the World with your Frontal Lobes.

These are just a few of the many available references that some readers may find useful in their own investigations beyond the scope of this book.

For those who wish to look into brain function in detail beyond the resources of the public library or the myriad of anything-goes information on the World Wide Web, a university or medical school library will generally offer public guest access to scientific and medical references, as well as to full medical and scientific journal collections.

Sarah J. Banks, Kamryn T. Eddy, Mike Angstadt, Pradeep J. Nathan, K. Luan Phan, "Amygdala-frontal connectivity during emotion regulation", Sarah J. Banks, Social Cognitive and Affective Neuroscience, Vol. 2, Is. 4, p. 303, (2007)

Mark G. Baxter, Elisabeth A. Murray, "The Amygdala and Reward", Nature Reviews Neuroscience, Vol. 3, Is. 7, p. 563-564, (2002)

Matt DeLisi, Zachary R. Umphress, Michael G. Vaughn "The Criminology of the Amygdala", *Criminal Justice and Behavior*, Vol. 36: p. 1241, (2009)

Rupa Gupta, Timothy R. Koscik, Antoine Bechara, Daniel Tranel, "The Amygdala and Decision-Making", *Neuropsychologia*, Vol. 49, p. 760-766, (2011)

Britta K. Hölzel, James Carmody, Karleyton C. Evans, Elizabeth A. Hoge, Jeffery A. Dusek, Lucas Morgan, Roger K. Pitman, and Sara W. Lazar. "Stress reduction correlates with structural changes in the amygdala." *Social Cognitive and Affective Neuroscience*, (2009)

Sara W Lazar, G, Gollub Bush, G.L. Fricchione, G. Khalsa G, Herbert Benson, "Functional brain mapping of the relaxation response and meditation", *NeuroReport*, 11: 1581-1585, (2000).

Joseph LeDoux, "Primer: The Amygdala", Current Biology, Vol. 17, Num. 20, p. 868-874, (2007)

Ekaterina Likhtik, Joe Guillaume Pelletier, Rony Paz, and Denis Pare "Prefrontal Control of the Amygdala", *The Journal of Neuroscience*, 25 (32): p. 7429-7437, (2005)

Elisabeth A. Murray, "The Amygdala, Reward and Emotion", *Trends in Cognitive Science*, Vol. 11, Num. 11, (2007)

James Olds, "Pleasure Centers in the Brain, Scientific American, (1956)

James Olds, Peter Milner, P., Positive Reinforcement Produced by Electrical Stimulation of Septal Area and Other Regions of Rat Brain", *Journal of Comparative and Physiological Psychology,* Vol. 47: p. 419-427, (1954)

"The Pleasure Seekers", *New Scientist*, October 11, (2003)
Rients Ritskes, "MRI Scanning during Zen Meditation", *Constructivism in the Human Sciences*, Vol. 8, p. 85-89, (2003)

H.B.M. Uylings, editor, "Cognition, Emotion and Autonomic Responses: The Integrative Role of The Prefrontal Cortex and Limbic Structures", *Progress In Brain Research*, Vol. 126, (2000)

INTERVIEW and WEB LINKS INDEX
(In alphabetical order)

THE AMAZING BRAIN ADVENTURE

NeilSlade.com
(TickleYourAmygdala.com)

The Brain Book and Music Store: neilslade.com/order.html

BookOfWands.com

EasyPaintYourCar.com

MyOwnPublishing.com

ErfieAndChloe.com

InkjetPrinterInfo.com

Julia Lu : JuliaPainting.com

Books by Neil Slade

The Frontal Lobes Supercharge

Brain Magic

The Secret of the Dormant Brain Lab
(The Book of Wands, Vol. 1)

The Book of Wands (Complete)

Have Fun Anti-Rules For Life, Learning, and Everything Else

Cosmic Conversations

Easy Paint Your Car

Easy Make A Kindle and Every Other Type of EBook